CW01263515

Land Warfare: Brassey's New Battlefield
Weapons Systems and Technology Series
Volume 10

Nuclear Weapons
Principles, Effects and Survivability

Land Warfare: Brassey's New Battlefield Weapons Systems and Technology Series

Executive Editor: Colonel R G Lee OBE, Former Military Director of Studies, Royal Military College of Science, Shrivenham, UK.
Editor-in-Chief: Professor Frank Hartley, Vice Chancellor, Cranfield Institute of Technology, UK.

The success of the first series on Battlefield Weapons Systems and Technology and the pace of advances in military technology have prompted Brassey's to produce a new Land Warfare series. This series updates subjects covered in the original series and also covers completely new areas. The new books are written for military personnel who wish to advance their professional knowledge. In addition, they are intended to aid anyone who is interested in the design, development and production of military equipment.

Volume 1	Guided Weapons – R G Lee
Volume 2	Explosives, Propellants and Pyrotechnics – A Bailey and S G Murray
Volume 3	Noise in the Military Environment – R F Powell and M R Forrest
Volume 4	Ammunition for the Land Battle – P R Courtney-Green
Volume 5	Communications and Information Systems for Battlefield Command and Control – M A Rice and A J Sammes
Volume 6	Military Helicopters – E J Everett-Heath, G M Moss, A W Mowat and K E Reid
Volume 7	Fighting Vehicles – Col. T W Terry, S Jackson, Col. C E S Ryley, B E Jones, and P J H Wormell
Volume 8	Surveillance and Target Acquisition Sytems – Col. A Figgures, Lt-Col. M McPherson, Prof. A L A Rogers, Dr P S Hall, T K Garland-Collins and J A Gould
Volume 9	Radar – Dr P S Hall
Volume 10	Nuclear Weapons: Principles, Effects and Survivability – Charles S Grace
Volume 11	Powering War: Modern Land Force Logistics – Colonel P D Foxton

Soldier dressed for NBC operations (Author)

Nuclear Weapons

Principles, Effects and Survivability

Charles S Grace RMCS
Visiting Senior Lecturer, Royal Military College of Science

BRASSEY'S (UK)
LONDON * NEW YORK

Copyright © 1994 Charles S. Grace

All Rights Reserved. No part of this publication may be reproduced, stored in a retrieval system or transmitted in any form or by any means: electronic, electrostatic, magnetic tape, mechanical photocopying, recording or otherwise, without permission in writing from the publishers.

First English edition 1994

UK editorial offices:
Brassey's, 165 Great Dover Street, London SE1 4YA

Orders:
Marston Book Services, PO Box 87, Oxford OX2 ODT

USA orders: Macmillan Publishing Company, Front and Brown Streets, Riverside, NJ 08075

Distributed in North America to booksellers and wholesalers by the Macmillan Publishing Company, NY 10022

Library of Congress Cataloging in Publication Data
Available

British Library Cataloging in Publication Data
A catalogue record for this book is
available from the British Library

0 08 040991 1 Hardcover
0 08 040992 X Flexicover

Charles S Grace has asserted his moral right to be identified as author of this work

Typeset by Florencetype Ltd, Kewstoke, Avon
Printed and bound in Great Britain by
Butler & Tanner Ltd, Frome and London

Acknowledgements

The material covered in this book was largely amassed during my many years of teaching and research at the Royal Military College of Science, Shrivenham, Discussions with colleagues, both civilian and military, over that period added much to my understanding of many topics. Here I must mention Lawrence McNaught (whose own work on this topic was a frequent source of ideas), Brian Dacre, Alan Avery, Mervyn Lovell, and, in particular, Ron West. I am most grateful to them all.

There are a number of others whom I must thank: Jean Mosley for her work on the diagrams, Andrew Farquhar and Ian Allen for their help with the illustrations, and Eve Turner and my wife for their typing. I also wish to thank the US Navy and Maxwell Laboratories for permission to use Figure 9.3. Finally, my special gratitude goes to Geoffrey Lee, the Executive Editor, for all his advice and guidance.

C. S. GRACE
Great Coxwell
Oxfordshire SN7 7L2
United Kingdom
February 1993

Contents

List of Tables	x
List of Figures	xi
Introduction	1
1 Nuclear Fundamentals	3
2 Nuclear Weapon Principles	14
3 Nuclear Weapon Outputs	25
4 Thermal Radiation and Effects	36
5 Blast and Effects	47
6 Nuclear Radiation and Effects	64
7 Electromagnetic Effects	91
8 Nuclear Survivability	106
9 Vulnerability Assessment	116
Bibliography	125
Glossary	127
Index	143

List of Tables

Table 1.1	Make-up of atoms	4
Table 1.2	Isotopes	6
Table 3.1	Fireball data	26
Table 3.2	Comparative outputs for fission weapon and ERW	32
Table 3.3	Effect of burst height on weapon outputs	34
Table 4.1	Correction factor for 1-km visibility	39
Table 4.2	Approximate thermal energies for ignition	42
Table 5.1	Shock wave velocity	48
Table 5.2	Effects of range and yield on blast-wave parameters	50
Table 5.3	Dynamic pressure and wind velocity	51
Table 5.4	Effect of reflection	54
Table 5.5	Ranges for blast damage	58
Table 5.6	Crater dimensions for surface bursts	61
Table 6.1	Early somatic effects of radiation	69
Table 6.2	Effects of beta radiation on skin	72
Table 6.3	Rate of delivery of immediate gamma ray dose	77
Table 6.4	Transmission factors for nuclear radiation	89
Table 7.1	EMP damage energies	100

List of Figures

Frontispiece   Soldier dressed for NBC operations

FIG. 1.1	The helium atom	4
FIG. 1.2	Shell structure of sodium atom	5
FIG. 1.3	The stability curve	9
FIG. 1.4	Radioactive decay	9
FIG. 1.5	Chain reaction in U-235	12
FIG. 2.1	Gun-type weapon	17
FIG. 2.2	The implosion weapon	18
FIG. 2.3	A high-yield weapon	23
FIG. 3.1	Variation of apparent fireball temperature with time; yield 20 kT	27
FIG. 3.2	Pressure variation in the blast wave	28
FIG. 3.3	Comparison of effects of ERW and fission weapon	33
FIG. 3.4	Time histories of weapon outputs; 1 km from 27-kT burst	34
FIG. 4.1	Variation of transmissivity with range, visibility 20 km	38
FIG. 4.2	Variation of thermal energy with range for 1 kT and 20 km visibility	39
FIG. 4.3	Variation of irradiance and total thermal energy with time	40
FIG. 4.4	RPL dosimeter reader before and after thermal irradiation	43
FIG. 5.1	Variation of static overpressure and dynamic pressure with time at a fixed weapon	49
FIG. 5.2	Formation of Mach stem	52
FIG. 5.3	Peak static overpressure (kPa) from a 1-kT burst	53
FIG. 5.4	Blast interaction with a box-shaped target	54
FIG. 5.5	Blast damage to vehicles	57
FIG. 5.6	Typical crater from an impact burst	60
FIG. 5.7	Surface cut-off in an underwater burst	62
FIG. 6.1	Neutron dose from a 1-kT weapon	74
FIG. 6.2	Immediate gamma ray dose	75
FIG. 6.3	Gamma dose rate 300 m from 1 kT	76
FIG. 6.4	Fallout dose rate variation with time	79

FIG. 6.5	The author using a portable dose rate meter for contamination measurement	85
FIG. 6.6	Rapid decontamination of vehicles	87
FIG. 6.7	Decontamination by scrubbing	88
FIG. 7.1	EMP from a surface burst	92
FIG. 7.2	Endo-atmospheric pulse	93
FIG. 7.3	Generation of exo-atmospheric EMP	94
FIG. 7.4	Role of magnetic field in Exo-atmospheric EMP generation	95
FIG. 7.5	Exo-atmospheric EMP: early time	96
FIG. 7.6	Area affected by exo-atmospheric EMP; heights of burst 100 km and 400 km	97
FIG. 7.7	(*a*) Spectrum comparison (*b*) Comparison with lightning	98
FIG. 7.8	Entry paths for EMP	101
FIG. 7.9	Global shield with waveguide below cutoff	102
FIG. 7.10	Hybrid protection	103
FIG. 8.1	Critical effects curve	109
FIG. 9.1	Blast tunnel	117
FIG. 9.2	Bounded wave EMP simulator	121
FIG. 9.3	Large, barge-mounted, radiating simulator for EMP assessment of ships	122

Introduction

Conventional munitions mostly rely for their effects on high explosives which, when detonated, give rise to a blast or shock wave. This causes damage to buildings and equipment, and injuries or death to man. In the past these effects have usually been fairly local – a single antitank missile, for example, can disable only one tank – but the recent introduction of cluster bombs and other submunitions means that this is no longer always the case.

Other conventional devices use the heat resulting from the combustion of unstable substances, such as napalm, but, again, the effects are confined to a comparatively small area. Despite this, these conventional munitions can produce devastating effects, particularly when they are associated with modern guidance techniques. Nevertheless, the weight and cost of the munitions needed to destroy a hostile tank force, or to disable all the aircraft and facilities on an airfield, are immense, as is the number of sorties required to deliver these munitions. Recent experience in the Gulf War bears this out.

By contrast, a nuclear weapon delivers its energy in three distinct forms – a blast, thermal radiation or heat, and nuclear radiation (unique to the nuclear weapon) – each of which may result in devastating effects on both man and equipment over a very wide area. There is a further, less well-known consequence of a nuclear detonation – the electromagnetic pulse ('EMP') – which does no harm to man but may have catastrophic effects on virtually all types of electronic devices and equipment.

Our experience of the combat use of nuclear weapons is confined to the two which were dropped on Japan in 1945 to bring World War II to an end. From subsequent studies of the targets (Hiroshima and Nagasaki) we have a fairly good understanding of the effects these weapons have on cities and their populations. By contrast, since they have never been used in anger against a military force, our direct knowledge of their effects on military equipment comes from nuclear weapon trials carried out before the signing of the partial nuclear test ban treaty by the major powers in 1963. For the effects on modern military systems we have to depend on various imperfect ways of simulating nuclear blast, thermal and nuclear radiation, and on theoretical studies.

As for military personnel, there is a great body of information about the effects on man of the blast and heat from conventional weapons. This gives us fairly good guidance on the likely effects of nuclear blast and heat on soldiers, sailors and airmen. There is, however, very little relevant experience as regards nuclear radiation. Perhaps even more importantly, there is no experience on which we may draw regarding the effects which a nuclear attack would have on the morale of troops in the field. It is fairly safe to assume that they would be detrimental and severe. Only troops who are well-trained and who have confidence in their equipment and their leaders, are likely to be psychologically fit enough to continue the battle after a series of nuclear strikes against them.

It might be thought that with the change in attitude shown by the Soviet Union in recent years, as evidenced, for example, by arms reduction treaties, the need for an understanding in military circles of the effects of nuclear weapons has diminished or disappeared. However, we must remember that nuclear weapons are no longer the sole preserve of a small handful of the larger nations (the USA, the former USSR, the UK, France and China). A significant number of other countries now have, or are capable of developing, nuclear weapons. If they do produce them, some of these nations are likely to be much less reluctant to make early use of them in a conflict than are the nations listed above. The study of nuclear weapon effects, therefore, remains relevant and essential.

The aim of this volume is to give the reader a broad understanding of nuclear weapons and their outputs, how these affect men and equipment, and the steps which may be taken to mitigate these effects. It does not seek to put the case for or against the development and employment of nuclear weapons. These topics have already been under more or less continuous debate by politicians, the military, and civilians of many countries for nearly half a century. This debate seems only likely to terminate should the day come when conventional weapons have been developed to the point where they can bring about a desired military aim as cheaply as nuclear weapons.

1.
Nuclear Fundamentals

Introduction

To understand the operation and effects of nuclear weapons, some knowledge of the nature and behaviour of the atom and its nucleus is needed. This chapter aims to provide these necessary fundamentals.

Atomic Structure

All the materials we encounter are built up from about 90 different natural elements. Each consists of atoms, which are the simplest entities able to have a permanent independent existence. They vary greatly in mass from the lightest (hydrogen) up to the heaviest (uranium), but all are built up from the same three basic constituents:

protons (p), which are positively charged;

neutrons (n), which have about the same mass as a proton but are uncharged; and

electrons, which have the same charge as protons but of opposite (negative) sign, and have only about 1/2000 of the mass of protons and neutrons.

Every element has a chemical symbol, e.g., U for uranium, and an 'atomic number' Z which is the number of protons in each atom. This is also the number of electrons, since atoms are normally electrically neutral, with equal numbers of protons and electrons. Every type of atom has a further number, the 'mass number' A, which is the number of heavy particles (protons plus neutrons, collectively called nucleons) in each atom. If we put X for its chemical symbol, a particular type of atom is represented as:

$^A_Z X$ e.g., uranium: $^{238}_{92}U$

This tells us that a uranium atom contains 92 protons, the same number of electrons, and 146 (i.e., 238–92) neutrons. Table 1.1 shows the make-up of a few examples.

Within the atom the protons and neutrons are crowded into a small, dense nucleus around which the electrons revolve in orbits, as Figure 1.1 shows. It is the electrostatic force of attraction between the negatively-charged electrons and the positive protons in the nucleus which provides the force holding the electrons in orbit round the nucleus. The overall size of the atom is very

TABLE 1.1
MAKE-UP OF ATOMS

Symbol	Atom	Protons	Neutrons	Electrons
$^{1}_{1}H$	Hydrogen	1	0	1
$^{4}_{2}He$	Helium	2	2	2
$^{12}_{6}C$	Carbon	6	6	6
$^{35}_{17}Cl$	Chlorine	17	18	17
$^{75}_{33}As$	Arsenic	33	42	33
$^{238}_{92}U$	Uranium	92	146	92

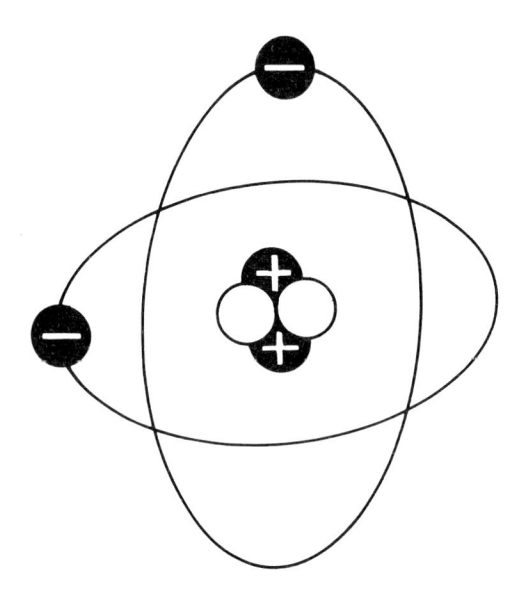

FIG. 1.1 The helium atom

small – about 10^{-10} metre; the mass varies from about 10^{-27} kilogram for the lightest (hydrogen) to about 240 times this for the heaviest. Because the nucleus contains all the heavy constituents of the atom (the protons and neutrons), most of the atom's mass is contained in it, but its diameter is only about 1/10,000 of the atomic diameter. The helium nucleus shown in the Figure is vastly exaggerated; to be in scale its diameter should be less than 0.01 mm. Nuclei therefore present rather small targets.

The chemical behaviour of an element – that is, how its atoms combine with atoms of other elements to form molecules of chemical compounds – is determined by the number and the arrangement of its electrons in their orbits. Indeed, the formation of a molecule involves a sharing or redistribution of electrons between them; the nuclei play no part in this.

Energy Units

Remembering the small size of each atom, it is hardly surprising to find that the energy evolved when atoms combine to form a molecule is also small – usually about 10^{-19} joule. The joule (J) is therefore much too large a unit to use when discussing atomic and nuclear phenomena. The unit customarily used is the electron-volt (eV), which is the energy acquired by an electron when it moves through a potential difference of one volt. It is equal to about 1.6×10^{-19} J. Within the nucleus we shall find the million electron-volt (MeV) more appropriate. A typical chemical reaction such as the burning of coal, in which an atom of carbon combines with two atoms of oxygen to form a molecule of carbon dioxide ($C + O_2 \rightarrow CO_2$) releases around 1 eV.

The Electrons

The atomic electrons occupy a series of orbits or shells round the nucleus. The innermost, called the K shell, can take only two electrons. The next, the L shell, accommodates up to eight, and several further shells may exist. Normally the shells fill up from the innermost, so a sodium atom, which has 11 electrons, has 2 electrons in the K shell, 8 in the L shell, and only 1 in the M shell (Figure 1.2).

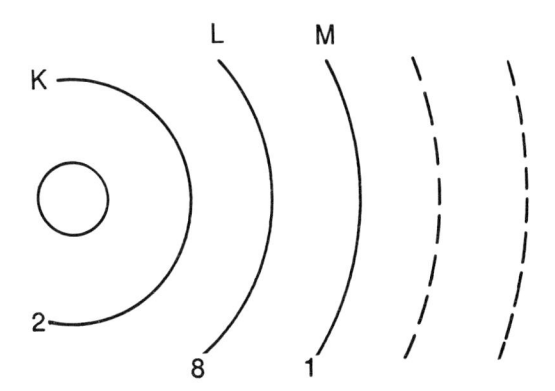

FIG. 1.2 Shell structure of sodium atom

If we heat or pass a current through sodium vapour the outermost electron may be given enough energy to move it out to a higher orbit (a process called excitation), or even to remove it completely from the atom, which is called ionization. The latter leaves the atom with a net positive charge; it is then described as a positive ion.

When an atom has been excited or ionized in this way, the electron involved quickly falls back to its normal orbit and in the process emits the energy it was given in the form of electromagnetic radiation. It appears as one or more 'packets' of electromagnetic radiation called photons. Their energy means that they usually lie in the visible light portion of the electromagnetic spectrum (which at shorter wavelengths also includes X- and gamma rays and, at longer wavelengths, radio waves). If the photon energy is between 1.5 and 3 eV it is ultraviolet light, while photons with energies below 1.5 eV lie in the infrared. This type of process is responsible for the light given out by a fluorescent tube or a sodium street lamp.

Similar effects occur with the inner electrons, but the energies involved are greater, particularly in heavy atoms with many electrons. The photons emitted as the electron falls back now have energies in the range 10^4 to 10^5 eV (10 to 100 keV), and are what we know as X-rays. They are generated for medical or industrial use by accelerating electrons with a voltage of perhaps 250 kV and directing them on to a target made of a metal such as tungsten ($Z = 74$). Inner electrons in the tungsten atoms are raised to higher orbits when the bombarding electrons collide with them; the X-rays are emitted as they fall back.

Isotopes

From what was said earlier we might conclude that there are only 92 sorts of atom, ranging from hydrogen (atomic number $Z = 1$) to uranium ($Z = 92$). This is not the case, since most elements occur in more than one form called isotopes. Chlorine is an example: about 75 per cent of chlorine atoms are the ^{35}Cl listed in Table 1.1, but the rest are a different isotope, ^{37}Cl. Both, of course, have 17 protons and 17 electrons – if they did not they would not be chlorine. The difference lies in the number of neutrons in their nuclei; the first isotope has 18, and the other one 20.

Some elements occur in nature in only one isotopic form – sodium (^{23}Na) is an example – but most occur in two or more forms. Table 1.2 lists a number which will be important to us.

TABLE 1.2
ISOTOPES

Symbol	Name	Number of Neutrons	Abundance (per cent)
$^{1}_{1}$H	Hydrogen	0	99.985
$^{2}_{1}$H	Deuterium	1	0.015
$^{3}_{1}$H	Tritium	2	–
$^{235}_{92}$U	Uranium-235	143	0.7
$^{238}_{92}$U	Uranium-238	146	99.3

Hydrogen is unique in that its isotopes are given the individual names deuterium and tritium (the latter does not occur in nature). In all other cases, such as uranium, the isotopes are usually referred to as uranium-235 (or U-235) and uranium-238 (U-238).

Nuclear Forces

Most of the interactions we experience in the everyday world can be explained by one of two forces – electrostatic forces and gravitation. The gravitational force holds the earth in orbit round the sun, but it is a comparatively weak force, and it is electrostatic forces which keep electrons in orbit round the nucleus (and, incidentally, hold atoms together in solids to give them their strength). A problem arises with the nucleus though: what keeps the protons and neutrons together within it? Electrostatic forces tend to eject the protons, which, of course, must electrically repel one another, and gravitation is far too weak to counterbalance this.

The answer lies in a third force, the nuclear force, which only acts between nucleons, i.e., particles in the nucleus. It is very short-range, unlike the other two forces. It only acts between neighbouring nucleons (which explains why we are not aware of it in everyday life), and is always attractive, providing what might be called a nuclear glue. Table 1.1 shows us that in a light atom such as carbon (six protons), an equal number of neutrons is needed to provide sufficient nuclear force to hold the nucleons together. When we get to heavier atoms, such as arsenic with 33 protons, the number of neutrons necessary is 42, or about 1.27 neutrons to each proton. For the heaviest, such as uranium, the ratio has to rise to about 1.6 if the nuclear force is to overcome the repulsive forces between the protons and so permit the nucleus to exist. Figure 1.3 illustrates this trend.

Binding Energy and Stability

Very accurate measurement of the mass of every type of atom, coupled with the Einstein mass-energy relation $E = mc^2$, leads us to a quantity called the binding energy per nucleon for every type of atom. This is simply the amount of energy needed on average to abstract a nucleon from the nucleus of a particular sort of atom. It is a valuable measure of stability – the greater the energy needed, the more stable is that particular nucleus. The numbers tend to be large (usually several MeV), which tells us that the energy evolved in any nuclear transformation is likely to be many orders of magnitude greater than the 1 eV or so usually released in the chemical reactions we discussed earlier.

The lightest atoms have a rather low binding energy – for deuterium (hydrogen-2) it is 1.1 MeV per nucleon – but as the atomic number increases it rises rapidly, reaching 7.7 MeV by the time we get to carbon-12. Thereafter it rises more slowly to a maximum for middle-weight atoms, then steadily falls off as we go on towards uranium. For example, for the atom tin-120 it is 8.5 MeV per nucleon, but falls to 7.6 MeV when we reach uranium.

From these facts we may conclude that if two very light nuclei, such as hydrogen isotopes, were to come close to each other, they should tend to fuse together, releasing a large amount of energy in the process. This does not happen spontaneously in practice because, as they approach, their protons mutually repel one another and stop them fusing together. They create what is called a 'potential barrier'. We may also conclude that heavy nuclei should try to achieve greater stability by spontaneously splitting into two lighter fragments. Again a potential barrier inhibits this, but, as we shall see shortly, in the case of the heaviest atoms it does not prevent it completely (though it does stop it from happening instantaneously).

We noted earlier that, for a nucleus with a particular mass number A (i.e., total number of nucleons), there is an optimum ratio of neutrons to protons; the case of arsenic-75, with 33 protons and 42 neutrons, was mentioned. If we look at other nuclei with 75 nucleons, such as germanium-75 (32p, 43n) or selenium-75 (34p, 41n), or any other possibility, we find that all have a lower binding energy per nucleon than that of arsenic-75, which must therefore be the most stable. We might therefore expect there to be a tendency for nuclei with either too many neutrons (such as germanium-75) or too many protons (such as arsenic-75) to change in some way which would increase their binding energy, and so their stability. We shall see that in fact they do. Figure 1.3 illustrates these tendencies we have been discussing; it shows how nuclear size and composition affect stability.

Radioactive Decay

Some of the unstable nuclei we have been discussing do manage to achieve greater stability by one of two types of spontaneous nuclear transformation called 'radioactive decay'. The first is alpha decay: it is the ejection of a composite particle comprising two protons and two neutrons, called an alpha particle. It is, in fact, identical with a helium-4 nucleus. All the elements with atomic number Z above 82 (lead) have isotopes which show alpha decay. U-235 is an example; the particles it loses in the transformation mean that the product has a mass number of 231 and an atomic number of 90 (thorium). Most of the several MeV of energy released is carried off by the alpha particle, but sometimes energetic photons of electromagnetic radiation – gamma rays – are emitted as well.

The second decay mode, called beta decay, happens in nuclei with a less than optimum ratio of protons to neutrons, i.e., those which lie off the stability curve shown in Figure 1.3. Unlike alpha decay, which is confined to heavy atoms, this process may occur in unstable nuclei of any size. There are actually two different modes of beta decay, which occur in either neutron-rich or proton-rich nuclei, but only the first concerns us. In neutron-rich nuclei a neutron spontaneously changes into a proton, which stays behind in the

Nuclear Fundamentals

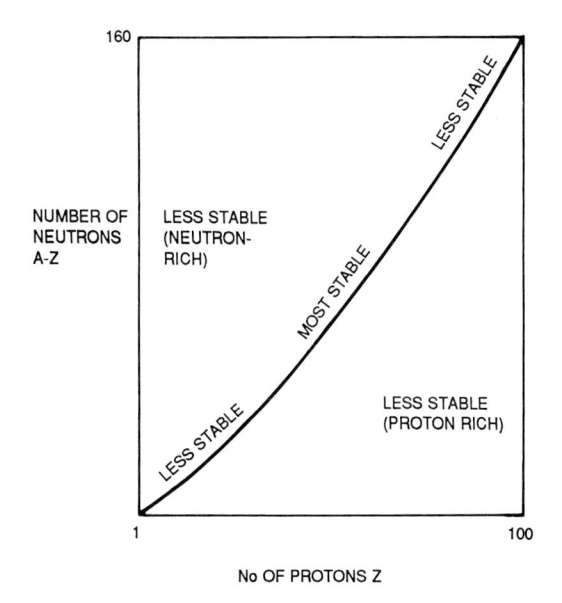

FIG. 1.3 The stability curve

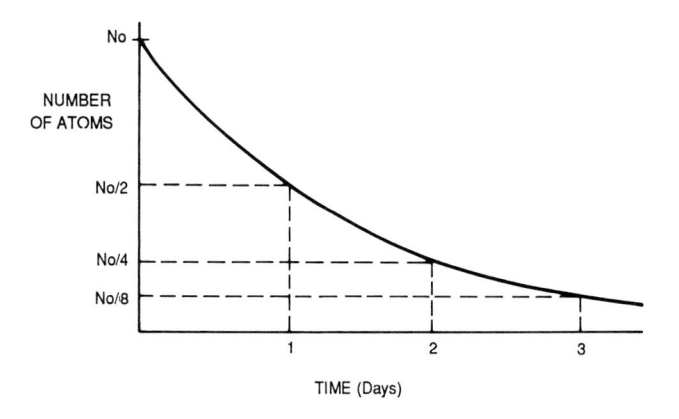

FIG. 1.4 Radioactive decay

nucleus, and an electron, for historical reasons called a beta particle, which is ejected at high speed. Energetic gamma rays usually accompany the disintegration. Just as alpha decay moves a nucleus down the stability curve towards greater stability, so beta decay moves an atom, which initially lies above it, diagonally downwards towards stability.

Half-life

Every type of radioactive atom decays at its own characteristic rate, called its half-life; it is the time taken for half the atoms in a sample to decay. An important feature is that the probability of a particular atom decaying is not influenced by how long it has already survived. A consequence is that if a radioactive isotope ('radioisotope') has a half-life of one day, then after one day half the sample will survive. During the second day half the survivors will decay, leaving one quarter of the original number, and so on. Figure 1.4 shows this behaviour. The decay is therefore described by the following relation:

$$N_t = N_0/2^{(t/t_{\frac{1}{2}})}$$

where N_0 is the original number of atoms, N_t is the number left after time t, and $t_{\frac{1}{2}}$ is the half-life.

A quantity of radioactive material is usually described not in terms of the number of atoms (N_0) it contains, but rather in terms of the number of atoms decaying per second, dN/dt, called its activity, which is much easier to measure. The activity falls with time in exactly the same way as the number of atoms. Half-lives of radioisotopes vary enormously, from very short (less than 10^{-6} seconds) to very long (greater than 10^{15} years). As we might expect, the most unstable atoms tend to have the shortest half-lives.

Natural Radioactivity

Virtually all the radioisotopes which occur naturally are found among the heaviest elements between lead ($Z = 82$) and uranium ($Z = 92$). All are daughters of three radioactive but long-lived isotopes, uranium-235, uranium-238 and thorium-232, which have half-lives comparable with the age of the earth (about 10^9 years). These all undergo a series of alpha and beta decays, ultimately becoming stable isotopes of lead. So we see why radium, with a half-life of only 1,600 years, is found on earth – it is continually being produced by the decay of its great-great-great grandparent uranium-238.

This explains why no elements beyond uranium, known as 'transuranics', occur naturally. They can exist, indeed one (plutonium) has been made in tonne quantities, as we shall see in Chapter 2, but none has a half-life long enough to permit detectable quantities of them still to exist in the earth's crust.

Induced Nuclear Transformations

We have just been looking at radioactive decay, which is a spontaneous nuclear transformation or reaction. Early this century it was found that when alpha particles from radioactive materials, such as radium, encountered atoms, most would expend their energy ejecting electrons from these atoms. Nevertheless a few – perhaps one in a million – would strike one of the very small targets which nuclei present and induce a reaction. Subsequently it was found that other fast, charged particles, *e.g*, protons (hydrogen nuclei) and deuterons (deuterium nuclei) produced by accelerators such as cyclotrons, could also induce nuclear reactions, but also infrequently. These reactions, though inefficient, turn out to be rather useful, for the product is usually radioactive. Very few of the elements have naturally-occurring radioisotopes, but by the suitable choice of a target element one may use induced reactions to provide artificial radioisotopes of virtually all the elements. Many have important applications in industry and medicine as tracers.

There is one reaction induced by charged particles which is particularly relevant to us; it involves the two hydrogen isotopes deuterium and tritium:

$$^{2}_{1}H + ^{3}_{1}H \rightarrow ^{4}_{2}He + ^{1}_{0}n + 17.6 \text{ Mev}$$

It may be induced by using an accelerator to project deuterons at a target containing tritium. It is an example of the fusion of light nuclei mentioned earlier, and is notable for its large energy release, over 14 MeV of which is carried off by the neutron. However, as a source of nuclear energy, this approach is hardly a winner. Only perhaps one in 10^5 of the deuterons we accelerate approaches close enough to a tritium nucleus to fuse with it. Consequently much more energy is expended in operating the accelerator than is released in the few reactions which take place. We shall see in Chapter 2 that there is an alternative approach.

Neutron-Induced Reactions

Neutrons are emitted in many of the nuclear reactions we have been discussing. During the 1930s it was found that neutrons are themselves very efficient at inducing nuclear reactions. Indeed, virtually every neutron so produced is ultimately captured by a nucleus. The reason is that a neutron has no electric charge, so it does not experience a repulsive force as it approaches a target nucleus in the way that an alpha particle, deuteron or proton does. Most neutrons emitted in reactions are fast-moving – perhaps 10^7 metres/second – corresponding to a kinetic energy of around 1 MeV. However, they are much more likely to be captured if they are slow-moving. It is easy to slow them down by allowing them to collide with certain lightweight atoms which show little tendency to capture them. The most suitable are hydrogen-2 (deuterium), carbon and hydrogen-1. The hydrogen isotopes are used in the form of water, the carbon as graphite. When used for this purpose they are called 'moderators'; they slow neutrons down until their energy is the same as that of the atoms of the moderator itself. They are then in thermal equilibrium with the moderator and are known as 'thermal neutrons'. By this time their velocity is about 2,000 metres/second and their kinetic energy about 0.025 eV.

Many isotopes capture neutrons, particularly those of thermal energy, very readily. In these reactions there is usually no particle ejected; the energy release is in the form of several energetic gamma-ray photons. An example is manganese:

$$^{55}_{25}\text{Mn} + ^{1}_{0}\text{n} \rightarrow ^{56}_{25}\text{Mn} + \text{gamma rays (7.3 Mev)}$$

The product has one more neutron than the original nucleus had, so is often radioactive. Being neutron-rich, its decay is by beta decay. This is the case with manganese-56; it emits 3.7 MeV of beta and gamma energy and has a half-life of 2.5 hours. The greater proportion of the radioisotopes in general use are produced by neutron reactions, with the use of nuclear reactors to provide the neutrons.

Nuclear Fission

One isotope of uranium, U-235, behaves in a unique way with neutrons. On capturing a neutron the atoms splits into two roughly equal-sized fragments (called fission products) with the release of a number of neutrons, about 2.5 on average. There is also a massive energy release of about 200 MeV – which incidentally, we can predict from what we know of the binding energies per nucleon of heavy nuclei such as U-235 (7.6 MeV per nucleon) and middle-weight elements such as the fission products (8.5 MeV per nucleon). U-235 is

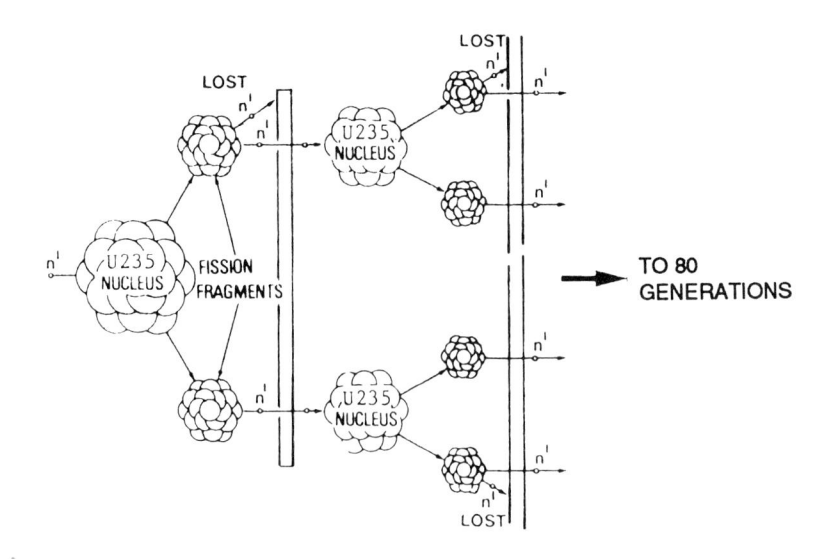

FIG. 1.5 Chain reaction in U-235

consequently described as a 'fissile' material. The fission products have about the same ratio of neutrons to protons – about 1.6 – as uranium. However, for stability, middle-weight elements require a ratio of only about 1.25. For this reason the fission products are highly unstable and suffer several beta decays before they finally achieve stability.

The particularly interesting features of fission are the large amounts of energy evolved and the release of several neutrons, which could go on to induce further fissions, so giving us a 'chain reaction' (Figure 1.5). There is nothing new about chain reactions; we have known and used them for thousands of years. The burning of coal and wood are chemical chain reactions, maintained not by neutrons but by heat. The discovery of nuclear fission in 1938 opened up for the first time the possibility of a nuclear chain reaction. We discuss its exploitation in the next chapter.

2.
Nuclear Weapon Principles

Introduction

In the first chapter we looked at the neutron-induced fission of uranium-235. We now go on to show how a fission chain reaction is exploited in a nuclear weapon, and how an alternative fissile material (plutonium) can be produced and used. We shall also see how the fusion of light elements may be used to improve weapon efficiency and to greatly increase the energy release.

Weapon Principles

Production of Fissile Material

Fission neutrons in natural uranium are much less likely to induce fission in the small amount of U-235 present than to be captured (without causing fission) in the much more abundant U-238. The first step is therefore enrichment of the uranium; this is the process of increasing the U-235 content from the normal 0.7 per cent to about 90 per cent. It cannot be done by chemistry since the two isotopes are chemically identical. Instead one must use physical methods which exploit the rather small difference in mass of the atoms of the two isotopes.

Enrichment methods which we can use include:

    electromagnetic separation, using instruments such as scaled-up mass spectrographs;

    gaseous diffusion (gaseous compounds of the two isotopes diffuse through a porous membrane at slightly different rates); and

    centrifuge methods which exploit their density difference.

The by-product of all these is uranium from which much of the U-235 has been removed. It is called 'depleted' uranium; we shall see that it, too, has a role in weapons.

Shape and Size

In a mass of fissile material, a neutron from fission has three possible fates:

    it can induce a further U-235 nucleus to fission;

Nuclear Weapon Principles

it may be captured by some other material, such as residual U-238; or

it may escape through the surface and be lost.

We can maximize the probability that it will induce a further fission by using highly-enriched uranium, and, secondly, by configuring the material so that, for its volume, it has the smallest possible surface area through which neutrons might escape. The shape which achieves this is a sphere.

Another factor which influences the creation of a chain reaction is size. In a very small sphere, only a centimetre or so in diameter, fission neutrons will be more likely to escape than to cause further fissions. With a larger sphere, the fraction which escape will be less. A useful concept here is that of the multiplication factor. As we shall see, a nuclear explosion is initiated by injecting neutrons into fissile material. These induce a certain number of fissions n_1 – the first generation of fissions. The neutrons from these then induce n_2 fissions in a second generation, and so on. The ratio of the number of fissions in successive generations is called the multiplication factor k for the system, i.e.,

$$k = n_2/n_1 = n_3/n_2, \text{ etc.}$$

If our fissile material has k less than 1.0 a chain reaction will not persist since there will be fewer fissions in each successive generation; it is said to be 'subcritical'. If k is exactly 1.0 there is the same number of fissions in each generation, and energy is released at a constant rate; it is then said to be 'critical'. The mass of fissile material giving us k equal to 1.0 is called the 'critical mass'. This is the condition we have in a nuclear reactor running at a steady power, but it does not give us an explosion. For this we must have k greater than 1.0, so that there are more fissions in each successive generation. The greatest value which k can have is 2.5, which is the number of neutrons released on average by each fission.

Requirements for a Weapon

It is obvious that before detonation the fissile material must be configured so that it is safely subcritical; otherwise a stray neutron (and there are always a few about) could initiate the explosion. Detonation will involve changing the configuration by some means to a supercritical one, and simultaneously injecting a burst of neutrons to initiate a chain reaction. The total energy yield will then depend on how many generations of fissions take place before the chain reaction dies away. It is important to realise that a nuclear weapon differs from a conventional shell filled with HE, in which all of the HE decomposes when it is detonated. In a nuclear weapon, however, the chain reaction terminates when quite a small fraction of the uranium has been fissioned. We must look at the cause of this termination.

The energy release during the reaction obviously creates enormous temperatures and pressures in the fissile material, forcing it to expand. When this happens its density decreases, and the separation between adjacent nuclei increases. This gives the neutrons an increased chance of escaping, so the multiplication factor k falls. Sufficient expansion converts our supercritical mass into a subcritical one, and the chain reaction quickly terminates. Incidentally, one can show that there is a simple relation between the critical mass m_{crit} and the density:

$$m_{\text{crit}} = 1/(\text{density})^2$$

We must now look at the energy required from the weapon, and how long it will take for this to be released. For historical reasons nuclear weapon yield is expressed in kilotons (kT), one kiloton being the energy produced when 1,000 tons of the high explosive TNT is detonated. This is taken as being 10^{12} calories, or 4.186×10^{12} joules. A uranium atom weighs about 3.9×10^{-25} kg, and releases 200 MeV when it fissions. It follows that fission of all the atoms in 1 kg of U-235 releases close to 8.2×10^{13} joules, or 20 kT, which is a typical yield from a tactical weapon.

The time between the emission of a neutron and its induction of a fission is about 10 nanoseconds (10^{-8} sec). If we assume that once the weapon is in a supercritical state it has a multiplication factor k of 2.0 (i.e., each fission gives rise to two fissions in the next generation), it can be shown that we need 80 generations of fission to give a yield of 20 kT. This will take a little less than 1 microsecond (10^{-6} sec), but it is worth noting that if we assume that, after the eightieth generation, the reaction abruptly terminates, due to k falling to less than 1.0 as the weapon blows apart, then over 99 per cent of the energy is released in the last seven generations. If in a weapon the fissile material were unconstrained it would be blown apart well before the eightieth generation, terminating the chain reaction and resulting in a very low yield. We must therefore surround it by a heavy tamper to slow down the expansion and allow the required number of fission generations to take place. The tamper has another beneficial effect, since it tends to scatter back into the fissile material some of the neutrons which would otherwise escape. It therefore increases k quite considerably. For example, a bare sphere of U-235 must be 170 mm in diameter, weighing 49 kg, to be just critical ($k = 1.0$). If it is surrounded by a thick tamper of non-fissile U-238 the diameter needed for k to equal 1.0 falls to 118 mm and the mass to 17 kg.

Gun-Type Weapon

The simplest way of satisfying the requirements for a weapon is that used in the gun-type weapon shown in Figure 2.1. The U-235 is in the form of two hemispheres which are each subcritical, but which form a highly supercritical mass when brought together. The hemispheres are kept well apart within a tamper rather like a gun barrel; this might be made of depleted uranium.

Nuclear Weapon Principles

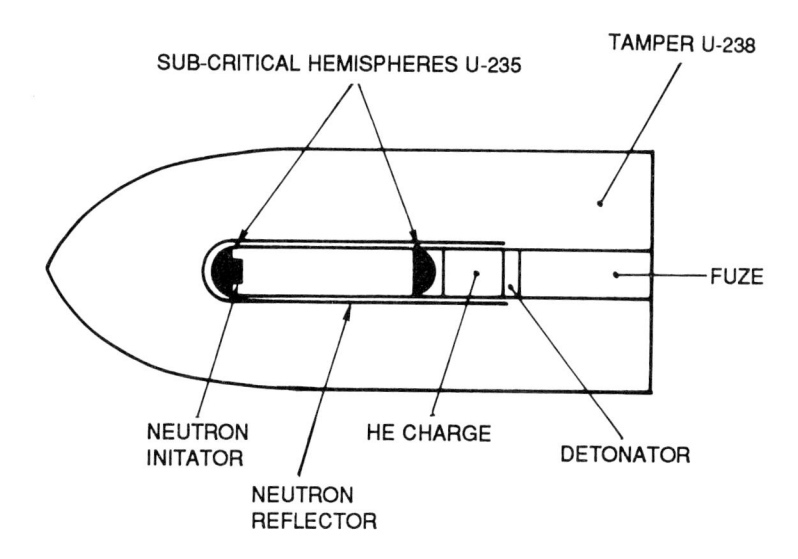

FIG. 2.1 Gun-type weapon

The nuclear explosion is initiated by detonating a high explosive charge behind one of the hemispheres, which accelerates rapidly down the barrel toward the other. At the instant they meet, a burst of neutrons is injected to initiate the chain reaction. The neutrons are provided by the fusion reaction we looked at in Chapter 1; here a miniature accelerator fires a burst of deuterons at a tritium target.

It is interesting to look at the precision which is needed in timing. The velocity of the moving hemisphere as it nears the fixed one might be about 1 km/sec, i.e., 1 mm per microsecond. Therefore the whole 80 generations will occur in a time during which the moving hemisphere travels less than 1 mm. If the neutrons were injected while the hemispheres were still about a centimetre apart (and k only 1.3 or so, instead of 2.0), the weapon would blow apart after perhaps two microseconds. During this the hemispheres move closer by only a millimetre or two, so k would never get to more than perhaps 1.5. This would result in a yield far below the design value, described as a 'fizzle'. It would be a quite unacceptable state of affairs for a weapon, the yield of which must be accurately predictable.

The gun-type weapon has one important limitation: its inefficient use of expensive fissile material. We saw earlier that, even with a thick tamper, 17 kg of U-235 are needed to achieve criticality. One might guess that some 30 kg would be needed to achieve a multiplication factor of 2.0 after assembly. If the weapon yield is 20 kT, only about 3 per cent is fissioned, the rest being wasted. A low-yield weapon would be even more inefficient. This is one reason why this basically simple design has long been superseded by more efficient alternatives.

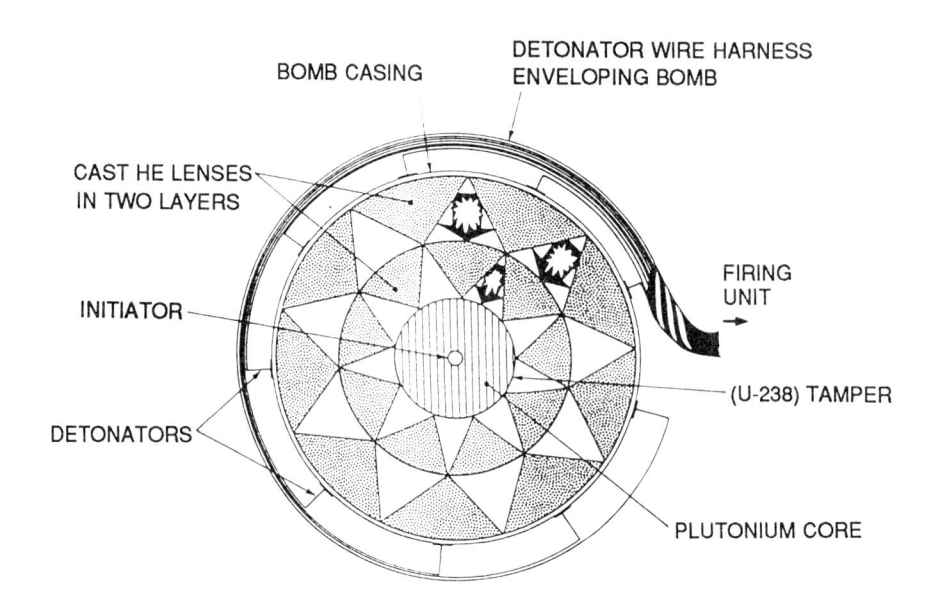

FIG. 2.2 The implosion weapon

Implosion Weapon

We saw earlier that decreasing the density of the fissile material increases the critical mass, and so diminishes the multiplication factor of a given mass. The converse also holds; that is, if the density is increased, k increases. This is the principle of the implosion weapon shown in Figure 2.2 and employed in modern designs of weapon. The fissile material is in the form of a small subcritical sphere. It is surrounded by a tamper which contains the fissile material for the required number of generations and also conserves neutrons which might otherwise escape. Outside this is high explosive, which is detonated simultaneously at a number of points on the exterior to produce a highly symmetrical, inwardly-travelling shock wave ('implosion'). This compresses the fissile core to around two to three times its normal density, so reducing the mass needed for criticality to between a quarter and one-eighth of the mass needed at normal density, and raising the multiplication factor from less than 1.0 to about 2. At the instant of maximum compression, a burst of neutrons is injected to initiate the chain reaction. As in the previous type, the number of generations for which the tamper is able to contain the fissile material determines the yield.

Since the mass of fissile material must be less than the critical mass at normal density, this type of weapon needs less fissile material than does the equivalent gun-type and so is more efficient. However there are problems. All the detonators must fire simultaneously if the necessary symmetrical implosion is to result. Otherwise the designed yield may not be obtained. For the same reason the HE must be of high purity and consistency. The shape of the charge is crucial – it is not just a hollow sphere, but an extremely finely machined series of segments called lenses.

A further difficulty is that the principal materials involved differ greatly in density, and perhaps in other properties such as the thermal expansion coefficient. Great care is therefore needed to ensure, for example, that mechanical forces during delivery (such as those which a nuclear artillery shell might experience) will not cause distortion which could lead to a malfunction. It is perhaps relevant to note that whereas the gun-type weapon used at Hiroshima in 1945 had not been tested beforehand, it was thought necessary to test an implosion weapon in New Mexico in July 1945, before the use of a weapon of this type at Nagasaki.

Plutonium

U-235 is the only fissile material which occurs naturally. However, there is an alternative which can be produced in large quantities: the transuranic element plutonium-239, which has an atomic number of 94. To see how this is done we need to discuss nuclear reactors.

Nuclear Reactors

Earlier in this chapter it was said that to achieve a fission chain reaction in uranium, its U-235 content must be considerably increased; otherwise most of the fission neutrons are captured by the 99.3 per cent of U-238 in natural uranium. However if natural uranium is formed into cylindrical rods ('fuel rods') which are arranged in a lattice, and the space between the rods is filled with a moderator, such as graphite or heavy water, then fast neutrons from fissions within the rods are very likely to escape from the rods before they are captured by U-238. Once in the moderator they are slowed down and, provided the fuel rods are optimally spaced, the neutrons are likely to be down to thermal energy by the time they re-enter a fuel rod. They are then more likely to cause fission in the 0.7 per cent of U-235 than to be captured by the much more abundant U-238, so a chain reaction can occur.

This is the principle of the so-called 'thermal reactors' widely used for power production. They are so called, not because of the heat they produce, but because the fission chain reaction is maintained by thermal neutrons. Many thermal reactors such as pressurised water reactors (PWRs) have ordinary water as the moderator, but since it captures rather more neutrons than the alternatives, the uranium fuel needs slight enrichment to 2 or 3 per cent U-235.

Plutonium Production

In a thermal reactor, of the 2.5 neutrons released in each fission, one is required to induce a further fission, and about half the rest are captured by

U-238, forming U-239. This fairly quickly suffers two successive beta decays forming Pu-239, an alpha emitter with a half-life of 24,000 years. Therefore Pu-239 accumulates in the fuel rods of all thermal reactors. Indeed, many of the thermal reactors built around 40 years ago were primarily intended for plutonium production. The electricity they generated was merely a useful by-product. The plutonium is chemically separated from the uranium and the fission products during fuel reprocessing.

Use in Weapons

The importance of this is that Pu-239 is, like U-235, a fissile material. Indeed, it has a significant advantage; the average number of neutrons released when it fissions is about 3.0, compared with 2.5 for U-235. Consequently its critical mass is a good deal smaller. Earlier a critical mass of 17 kg was quoted for a sphere of U-235 at its normal density and surrounded by a thick tamper. For Pu-239 the corresponding figure is only 6.0 kg, corresponding to a sphere 90 mm in diameter. At twice normal density (which might be achieved by implosion) the critical mass is a quarter of the above, i.e., 1.5 kg. The advantages in size and efficiency which the use of plutonium brings with it are obvious.

There is a complication though. In the fuel rods of a reactor there is a build-up not only of Pu-239, but also of another isotope Pu-240, which tends to fission spontaneously with, of course, the emission of neutrons. Therefore in a plutonium weapon there is a risk of spontaneous fission causing initiation of the chain reaction before the optimum moment; that is, before maximum supercriticality is reached. It would result in a fizzle, i.e., a much reduced yield. This is more likely to occur in a gun-type weapon, since the change from a subcritical to a supercritical state takes longer in this design than it does in an implosion weapon. For this reason it is generally held that plutonium is unsuitable for use in gun-type weapons.

This is not the end of the story. In a reactor primarily operated for power production, generating costs are kept low by leaving the fuel rods in the reactor for a comparatively long time. However, the longer the fuel spends in the reactor the greater is the concentration of Pu-240 in the plutonium, so the greater is the spontaneous fission rate in it. It was once widely thought that this made plutonium from reactors run primarily for power production unsuitable for use in any weapons, even of the implosion type, because of uncertainty of the yield. Since then weapon design has advanced, and it is clear that experienced design teams can now produce reliable weapons using fissile material with a high Pu-240 content.

If a yield of the order of hundreds of kilotons is required there are practical (but not insuperable) difficulties, since such a weapon must contain many kilograms of fissile material, which has to be in a subcritical configuration before detonation. One solution might be to configure the material as a hollow sphere. This could contain a large mass of fissile material without

Nuclear Weapon Principles

going supercritical, provided its wall thickness was sufficiently small. However, if it were imploded it would rapidly change to a solid, dense, highly supercritical sphere, and a large yield would result.

However, there are also economic considerations, since both the enrichment of uranium and the production and separation of plutonium are very expensive processes. A much cheaper way of achieving high yields is described in the next section.

Thermonuclear Fusion

We saw in Chapter 1 that light nuclei release large amounts of energy when they fuse together, but this does not ordinarily happen spontaneously. The easiest fusion to initiate (which also has an unusually large energy release) is that of deuterium and tritium. It was pointed out that the use of an accelerator to project deuterons at a tritium target does give some fusions, but the accelerator consumes much more energy than is released by the fusions. Therefore this approach is not a useful basis for a weapon.

There is an alternative approach. If a mixture of the gases deuterium and tritium, at fairly low pressure, is electrically heated to a temperature of about a million degrees, the average kinetic energy of the nuclei will be about 1 keV, and in a time of a few seconds a significant number of fusions will occur. Fusion initiated by heat in this way is called 'thermonuclear fusion'; the JET experiment at Culham, near Oxford, and others elsewhere are aimed at developing this technique to the point where it can be used for civil power production. However, a thermonuclear reaction induced in this way is far too slow to serve as a basis for a weapon. A fission weapon, though, produces a much higher temperature of a few tens of million degrees, and a pressure of millions of atmospheres. If fusion reactants are exposed to these conditions their reaction rate is enormously increased, and a large proportion react in a time of the order of a microsecond, yielding several times as much energy per unit weight of reactants than would be obtained by the fission of heavy nuclei. This is the basis of thermonuclear weapons; the fusion reaction is triggered by a fission device.

There are two problems. First, both reactants are low-density gases, so it is not easy to incorporate kilogram quantities into a weapon. Secondly, although deuterium occurs naturally and is fairly easily separated from hydrogen-1, tritium is not found in nature and is also radioactive with a half-life of about 12 years. A simple solution exists to both these difficulties: the use of lithium deuteride as the fuel for the fusion stage. This is simply lithium hydride containing H-2 instead of the usual H-1; it is a waxy solid. The fission trigger yields many more neutrons than are needed to maintain the chain reaction. If some strike the lithium deuteride they initiate a reaction which yields tritium:

$$^{6}_{3}\text{Li} + ^{1}_{0}\text{n} \rightarrow ^{4}_{2}\text{He} + ^{3}_{1}\text{H} + 4.8 \text{ MeV}$$

Under the prevailing conditions of temperature and pressure the tritium from this reaction readily fuses with the deuterium from the lithium deuteride.

Unlike the products of fission, which are highly radioactive, as we shall see, the product of thermonuclear fusion is the stable nucleus helium-4. It follows that a weapon obtaining most of its energy from fusion would produce much less radioactivity; it could be described as a 'clean weapon'.

Weapons using Fusion

Thermonuclear fusion is employed in three quite distinct ways in weapons. The rest of this chapter looks at each of these in turn.

Boosted Fission Weapons

The efficiency of a fission weapon is markedly increased if a small amount of fusion fuel is incorporated in it. The fusion stage does not contribute much directly to the energy yield, but its high energy (14 meV) neutrons increase the speed of the build-up of the fission chain reaction, so that many more fissions occur before expansion terminates the fission chain reaction. This process is called 'boosting'.

High-Yield Weapons

A high-yield weapon can be made with a small fission trigger which initiates a much larger fusion stage providing most of the energy yield. Devices like this have been tested, but there is a cheaper approach, the principles of which are shown in Figure 2.3. This uses a depleted uranium tamper round a thermonuclear weapon. Although the principal constituent of the tamper, U-238, does not fission when struck by fission neutrons (with a most probable energy below 1 MeV), the much more energetic neutrons from fusion readily induce it to fission. So, in a weapon like that shown a large number of fissions occur in the depleted uranium tamper. Most high-yield weapons are of this type. Although they are known as 'thermonuclear' weapons or 'hydrogen bombs', about half the energy originates in fission, mainly in the tamper. They are therefore 'dirty weapons', producing much radioactivity.

It should be mentioned here than an actual high-yield weapon would be rather different from that illustrated in Figure 2.3, which might not work very efficiently, if at all. Interchanging the positions of the fission trigger and the fusion material would be better, since the energy released by the fission stage would tend to compress the fusion material, so increasing the speed and efficiency of the thermonuclear reaction. However, if we wished to incorporate a large quantity of fusion material, the amount of fissile material needed for the trigger would rise dramatically and expensively.

We shall see in the next chapter that the fission stage releases not only a

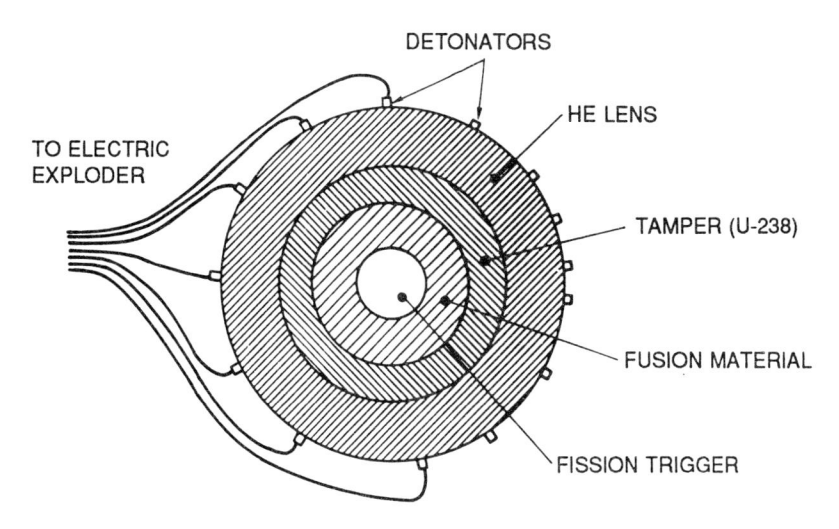

FIG. 2.3 A high-yield weapon

large number of neutrons, but also an intense flux of soft X-rays. Real high-yield weapons must be configured so that the X-rays are directed on to the fusion material in such a way that they heat and compress it as effectively as possible. This requires a rather different geometry.

Enhanced Radiation Weapons

It is possible to produce comparatively low-yield weapons with only a small fission trigger to initiate a fusion stage. If it is designed so that the nuclear reactions proceed as fast as possible, the tamper need not be very thick, and a large proportion of the energetic fusion neutrons will escape. As we shall see in the next chapter, this greatly increases the nuclear radiation output compared with that from a fission weapon of the same yield, at the expense of heat and blast. Weapons of this type are called 'enhanced radiation' weapons, or, popularly, 'neutron bombs' (though delivery as an artillery shell or missile warhead is much more likely than release from an aircraft).

Conclusion

We conclude with some brief data to summarise the advances which have been made since the first nuclear explosions in 1945. The gun-type weapon used at Hiroshima contained about 42 kg of uranium (80 per cent U-235). It yielded 12.5 kT, so its efficiency was less than 2 per cent. Its length was 3.2 m, its diameter 0.75 m, and it weighed 4.4 tonnes.

The first generation implosion weapon used at Nagasaki contained some 2 tonnes of HE in 96 separate components; they formed a spherical assembly about 1.5 m in diameter. Detonation at 32 points gave a yield of about 20 kT.

By contrast, modern designs can be accommodated in a 155-mm artillery shell (though they would yield only 1 kT or so). The largest fission device ever tested (USA, 1952) yielded 500 kT.

The first true thermonuclear device, tested by the USA in 1952, employed liquid deuterium and tritium as the fuels and yielded 10 MT, or the equivalent of 10 million tons of TNT. A great deal of cryogenic equipment was needed to maintain the fuels in the liquid state, so the complete device weighed 66 tonnes. Practical weapons using solid lithium deuteride as the fusion fuel, and capable of being delivered by aircraft or missile, were developed soon after. Nowadays several independent nuclear devices (MIRVs) can be carried by one missile; size is no longer the problem that it was with first-generation weapons.

3.
Nuclear Weapon Outputs

Introduction

The previous chapter explained the principles which underlie the design and operation of nuclear weapons. We now turn to the consequences of an enormous energy release within a comparatively small volume, and shall see that effects are produced which differ from those of conventional munitions in more than just scale.

There are four distinct outputs from a nuclear weapon; they are thermal radiation, blast, nuclear radiation, and electromagnetic pulse. For blast the main difference from that produced by ordinary high explosive is its magnitude. As for thermal radiation, conventional heat weapons, such as those using napalm, tend only to affect objects with which the weapon material makes contact; not much heat is radiated away to more distant targets. By contrast, thermal effects from a nuclear weapon are caused by radiated heat.

Nuclear radiation is, of course, unique to nuclear weapons, as is the (nuclear) electromagnetic pulse, normally called EMP or NEMP. It is a short intense pulse of radiofrequency energy, originally called 'radioflash', which is energetic enough to cause upsets or damage to most electronic and electrical systems. Unlike the other outputs, EMP does not originate in the weapon itself; it is a consequence of the interactions of nuclear radiation with the atmosphere. In this chapter we give only a brief introduction to the several weapon outputs and how they are influenced by burst location relative to the earth's surface. The outputs, and their effects on man and materials, are all covered in much more detail in later chapters.

Low Air Burst

Fireball Development

Here we consider the most likely scenario, which is a weapon burst in the atmosphere but fairly close to the ground. At the end of the chain reaction about 85 per cent of the energy is in the form of the kinetic energy of the fission products. The remaining 15 per cent is emitted as nuclear radiation, either immediately or at some later time; this need not concern us for the moment. The fission product kinetic energy is rapidly shared with the weapon debris, producing a temperature of several tens of million degrees and a pressure of several million atmospheres. Like any hot material, the debris immediately begins to radiate thermal energy. Most hot objects we experience have temperatures of a few hundred to, at most, a few thousand degrees. At these everyday temperatures the radiated thermal energy lies in the infrared and the visible part of the electromegnetic spectrum, with some ultraviolet as well from rather hotter objects such as molten metals or the surface of the sun (which is about 6000 degrees). The debris has a temperature several orders of magnitude higher, and so radiates intense thermal radiation; but it is dwarfed by radiated soft X-rays with a photon energy of several KeV. Most of these do not travel far: they are absorbed by the air, so spreading the energy over an ever-increasing volume around the point of detonation. This constitutes the nuclear fireball which expands to a size depending on the yield (Table 3.1). At the same time it rises, rapidly at first (about 100 metres/sec), and cools as it radiates away its energy. When it has cooled so much that it is no longer luminous the debris, initially in the form of a vapour, condenses to form a visible cloud. This continues to rise under buoyancy and other forces until it ultimately stabilises and spreads laterally at a height which depends on the yield – about 2.5 km for a 1 kT burst, or 20 km for 1 MT.

TABLE 3.1
FIREBALL DATA

Yield (kT)	Max. fireball radius (metres)	Time to t_{max} (seconds)
1	67	0.0417
20	222	0.156
1,000	1,062	0.871

Thermal Radiation

Based on our experience, we might expect the temperature T of the fireball and its rate of heat emission (which, by the Stefan-Boltzmann law, is proportional to T^4) both to fall steadily with time. As measured from the ground this is certainly not the case. Figure 3.1 shows the observed variation with time. Two factors cause this departure from our expectations. First, the nuclear radiations create absorbing compounds in the atmosphere. As a result, at 10^{-4} seconds after the explosion, when the true fireball temperature is close to 10^6 degrees, its temperature as measured from the ground is only about three thousand degrees.

As the absorbing compounds decay, the heat begins to get through to the ground, but another effect then occurs. The kinetic energy of the fission products is steadily being transferred to other weapon debris, leading to the generation of a shock front or blast wave which moves out through the

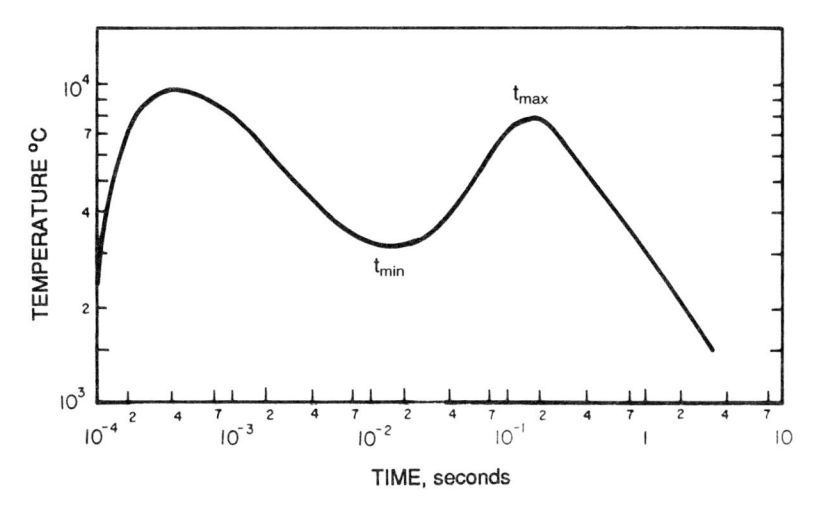

FIG. 3.1 Variation of apparent fireball temperature with time; yield 20 kT
(*US Department of Defense – Department of Energy*)

fireball into the surrounding air. Its pressure is extremely high and the air behind it is heated by compression, having the effect of making it opaque to the thermal radiation from the fireball. This blanketing effect of the shock wave is at its maximum at about 10^{-2} seconds, when the heat received on the ground is almost entirely from the incandescent air in the shock front. The time t_{min} in Figure 3.1 corresponds to this stage; it is called the 'thermal minimum'.

As the shock front spreads and weakens its masking effect decreases, and thermal radiation from the fireball reaches the ground unhindered for the first time. For the weapon in the Figure this occurs at about 0.16 seconds, which is called t_{max}. The fireball temperature measured from the ground, about 8,000 degrees, is now for the first time its actual temperature. Thereafter it cools steadily as it radiates its heat. The consequence of all this is that the thermal radiation reaches the ground in two pulses. Since the second pulse (after t_{min}) is much the longer, most of the heat reaching the ground – about 99 per cent – is in the second pulse. Indeed, for most purposes the effect of the first pulse may be ignored. The total length of the thermal pulse may be assumed to be the time in which 80 per cent of the thermal radiation is received. This is less than half a second for a 1 kT yield, but over 8 seconds for 1,000 kT. Weapon yield hardly affects the observed temperatures, but a weapon of higher yield produces a larger fireball which takes longer to cool. The time scale therefore depends on yield, as Table 3.1 shows.

Calculation of the total thermal energy from a weapon is simple since the weapon yield defines the total energy release, and for all types of weapon except enhanced radiation weapons the proportion appearing as heat is 35 per cent.

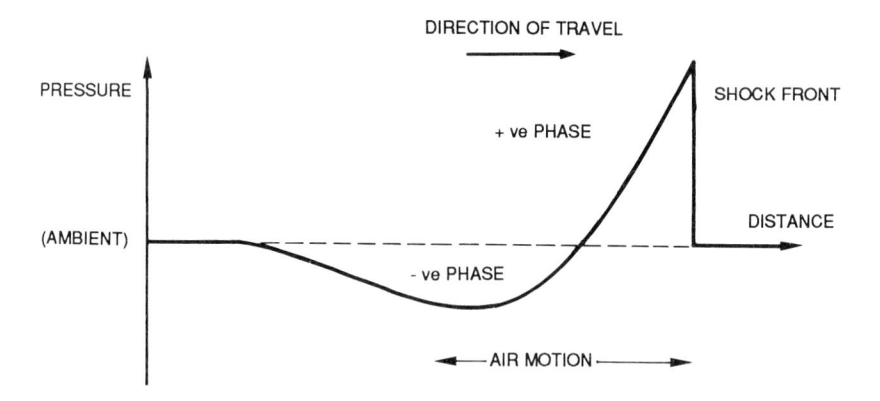

FIG. 3.2 Pressure variation in the blast wave

Blast

The origins of the blast wave were explained in the last section. It moves outward in all directions, weakening as it does so. Close to the point of burst its very high pressure causes its velocity to considerably exceed the normal velocity of sound (330 metres/second), but after it has travelled a kilometre or so its velocity falls to close to this figure, at least for low yields.

Its main features are shown in Figure 3.2. To the right of the shock front lies undisturbed air; to its left is a region of air pressure above the ambient ('positive overpressure') called the positive phase. Here the air density is above the normal value; indeed, it is this density increase which is responsible for the overpressure. Closer to ground zero (GZ, the point on the surface below the burst) is a region of underpressure, called the negative phase, in which the air density is below the normal value. This non-uniform density distribution implies that there has been a mass movement of air molecules, i.e., a wind is associated with the passage of the blast wave. During the positive phase it blows outward from GZ; in the negative phase its direction is reversed.

The overpressure produces effects on men, equipment and structures. The blast wind also exerts a pressure, called dynamic pressure, producing its own set of effects. For many military targets, it is this last which has the more serious effect. Usually the positive phase does the damage – the negative phase effects are much less significant. The ground has an important influence since it reflects the wave, thereby enhancing its effects.

Optimum Height of Burst

When a 1-kT weapon is employed against a relatively 'hard' target, such as a tank concentration or troops in foxholes, the blast damage radius is greatest

Nuclear Weapon Outputs

if the weapon is used as a low air burst, detonated 60 metres above the ground. For any other yield of W kT this 'optimum height of burst' is given by:

Optimum height = $60 \, W^{1/3}$ metres

For softer targets, such as aircraft, ships and buildings, the optimum burst height is greater.

Nuclear Radiation

About 15 per cent of the energy from all nuclear weapons (except enhanced radiation weapons) emerges as nuclear radiation. This may seem small when compared with the blast (50 per cent) and heat (35 per cent), but its effects may often be dominant. Man, for example, is much more severely affected by a given amount of energy delivered as nuclear radiation than he is by the same energy delivered as heat. A lethal radiation dose would, if delivered as heat instead of nuclear radiation, only raise the body temperature by about 0.001°C. It is about the energy he would receive by lying in the sun for half a second.

There are several sources of radiation associated with a weapon burst:

a. Neutrons from the fission (and, where applicable, fusion) reactions. Nearly all are emitted during the actual nuclear reactions, and, with velocities of up to 1/30 of that of light, arrive at ranges of interest from a few tens of microseconds up to a few tens of milliseconds after an explosion. However, a small number are emitted seconds or even minutes later. These 'delayed neutrons' make it imprudent to explode weapons close together in time and space. Delayed neutrons from the first could cause a malfunction of the second – an effect called 'fratricide'.

b. Gamma rays emitted during fission. With the same velocity as light, these travel a kilometre or so in around 3 microseconds.

c. Gamma rays originating when neutrons collide with or are captured by atoms in the weapon debris and the atmosphere. These begin to arrive with the fission gammas; the last of them arrive with the slowest neutrons.

d. Gamma rays and beta particles from the decay of the highly radioactive fission products. These are at their most active when just formed; thereafter their activity falls steadily with time. Unlike neutrons and gamma rays, some of which travel up to thousands of metres in air, beta particles have a range in air of a few metres at most. Beta particles originating from the debris in the fireball are therefore not a threat.

Immediate Nuclear Radiation

From an air burst weapon all the radiations in a, b and c above will reach a target in much less than a second. However, the fission product gamma rays are emitted over a very long period, but since the fireball containing the fission product rises rapidly, and since the rate of emission of gamma rays falls steadily with time, it is found that little radiation reaches the ground more than 10 seconds after the burst. It is usual to describe all radiation reaching the ground up to an arbitrary time of one minute as 'immediate nuclear radiation'.

For the air burst we are discussing the debris rises to form the radioactive cloud. As this drifts with the wind the fission products, now very small, solid particles, slowly settle towards the earth as fallout. Under these conditions individual particles take a time which ranges from days to months or years to reach the ground, so the fallout is worldwide. By the time they reach the ground they are so widely dispersed and their activity has decayed so much that they are not a military problem.

Residual Nuclear Radiation

This is nuclear radiation received more than one minute after an explosion. There are two separate contributions – fallout and neutron induced activity (NIA). We have seen that for an air burst the fallout is not significant, but there may be appreciable NIA since, as was explained in Chapter 1, the fate of all neutrons is to be captured. Neutrons which reach the ground are captured by nuclei in soil and equipment. The products of these reactions are usually radioactive, decaying by emission of beta particles and gamma rays. Whereas the beta particles in the immediate radiation are no threat to man and equipment, those from NIA do put man at risk over a roughly circular area round GZ. Certain elements, including manganese, aluminium, silicon, sodium and zinc, capture neutrons particularly readily. The intensity of NIA near soil or equipment will depend on the neutron dose and on the content of these elements in the soil or equipment. This activity decays fairly quickly in most cases, and is unlikely to be much of a problem after a day or two.

Electromagnetic Pulse

This is not discussed in detail here. It will suffice for the moment to say that, unless appropriate precautions are taken in the design and operation of electronic equipment, it is very likely to be damaged by EMP at ranges well beyond those at which it would be damaged by the other weapon outputs.

Nuclear Weapon Outputs

Other Burst Heights

Surface Burst

If a burst takes place on or very close to the ground a large crater is produced, but the blast wave is not greatly affected. The dust and debris thrown up absorb about half the thermal radiation, reducing the range of thermal effects. The immediate nuclear radiation is also affected, particularly for very high and very low yields. This will be discussed in Chapter 6.

It is the residual nuclear radiation, however, which is most influenced by a surface burst. More neutrons strike the ground so neutron induced activity is increased, particularly around GZ, but the major effect is on fallout. With a low air burst at 60 $W^{1/3}$ metres the fireball just fails to touch the ground and fallout is not militarily significant. But if the burst is lower than this, or is on the surface, a great deal of soil is drawn up into the fireball, providing a liberal supply of coarse particles on which the vaporised fission products condense. These fall from the radioactive cloud quite quickly, in a time of minutes or hours. The result is an area of intense local fallout, mainly downwind from GZ. Local fallout such as this, when it occurs, makes by far the biggest contribution to residual radiation. It may limit operations in the affected area for several days or more.

Underground Burst

The effects of a shallow underground burst are rather like those of a surface burst, but the greater the depth of the burst, the larger is the reduction in thermal and immediate nuclear radiation emerging into the atmosphere. More of the energy goes into the formation of a crater, and there is an intense shock wave in the ground. If the explosion is so deep that it does not break through the ground surface, then blast and thermal and nuclear radiation effects above ground are negligible, nearly all the energy going into the ground shock.

Underwater Burst

A burst below the sea surface produces no thermal nor EMP effects above or below the water, but both surface and submerged vessels may be damaged by the intense underwater shock wave. There will also be large waves on the water surface which may affect ships and shore installations. The first nuclear radiation received by a surface vessel will be from fission products and NIA in the column of water thrown up by the burst. As the column collapses most of the radioactivity settles in a doughnut-shaped mass of water round surface zero; it loses activity through decay and mixing.

Exo-atmospheric Burst

When a burst takes place within the atmosphere the X-rays referred to earlier are responsible for the growth of the fireball, the energy ultimately being radiated as heat. However, if the burst is above the atmosphere – say at a height greater than 35 km – then the low-energy X-rays, which comprise over 70 per cent of the energy output, are not converted to heat. Together with the gamma rays and neutrons they spread out, suffering little absorption as they do so. If they impinge on a target which is also above the atmosphere the consequences may be serious, even at distances of tens of kilometres. First, the deposition of this energy in a period measured in microseconds creates a severe mechanical shock which may be destructive. Secondly, the radiation gives rise to a phenomenon called 'system-generated electromagnetic pulse' (SGEMP) which may have serious effects on electronic circuitry.

As for effects on the ground, there is a pulse of thermal radiation, produced by the impact of the X-rays on the upper layers of the atmosphere. It consists of a single, short pulse, the only effects of which are on human observers, who may be temporarily dazzled. Blast and nuclear radiation effects on the ground are negligible, but there is an intense electromagnetic pulse over a very wide area.

Enhanced Radiation Weapons

Outputs

The information in the following chapters about weapon outputs, and how they are influenced by range and weapon yield, applies to weapons which obtain half or more of their energy from fission, i.e., to both tactical fission weapons and high-yield weapons. The enhanced radiation weapon, which has a rather thin tamper, and which obtains most of its energy from fusion, we must treat as a special case. Table 3.2 shows how the energy divides for fission weapons and enhanced radiation weapons (ERWs).

TABLE 3.2
COMPARATIVE OUTPUTS FOR FISSION WEAPON AND ERW

| | Percentage of energy | |
Output	Fission	ERW
thermal radiation	35	<20
blast radiation	50	<30
nuclear radiation		
immediate	5	> 50
fallout	10	negligible
NIA	slight	more

From the Table we see that it is reasonable to assume that an ERW yielding W kT produces blast and thermal outputs and effects similar to a fission weapon yielding $\frac{1}{2}W$ kT, but immediate nuclear radiation similar to a fission weapon yielding $10W$ kT. The fallout from an ERW is negligible since the product of the fusion reaction is stable helium, so the activity in the weapon debris mainly comes from the small fission trigger.

Comparison of Effects

Data in later chapters enable us to deduce that when low-yield weapons are used against well-equipped and well-trained troops, the immediate nuclear radiation is the ruling or governing effect – that is, it produces casualties out to a greater range from GZ than do blast or thermal radiation. So, if we are thinking of tactical weapons, an ERW is more efficient than a fission weapon in that it produces more casualties per kiloton by virtue of its greater output of nuclear radiation.

Perhaps its greatest advantage, though, is the reduction it gives in undesired but inevitable blast damage to houses, factories, forests, etc. (sometimes called 'collateral damage'). To illustrate this, suppose a 10 kT fission weapon were used against a tank concentration. Using data from Chapter 6 one can show that the crews of tanks up to 880 metres away would be disabled by nuclear radiation. However, there would be moderate to severe damage to houses out to 1800 metres. Had a 1-kT ERW been used instead, the effect on the tank crews would be the same, but the damage to houses would only extend out to 690 metres. Figure 3.3 illustrates this.

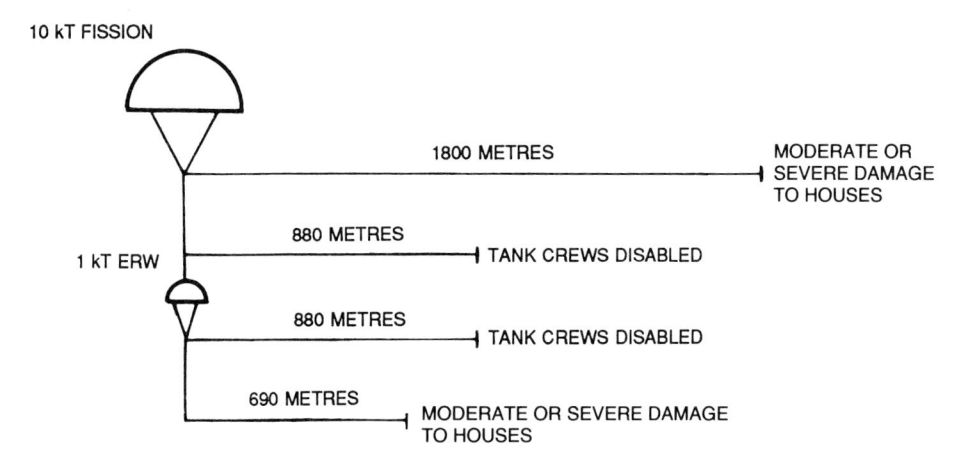

FIG. 3.3 Comparison of effects of ERW and fission weapon

More relevant is the area over which houses are damaged. Using the ERW, the area of blast damage is only 15 per cent of that for the fission weapon.

Indeed, by careful selection of yield and height of burst it would be possible to reduce the area of collateral blast damage from the ERW even further.

The principles of the ERW mean that in practice its yield is unlikely to exceed about 10 kT. Even if it were possible to produce ERWs with much higher yields there would be no point in doing so, since for higher yields either blast or thermal radiation becomes the ruling effect for all types of nuclear weapon. Therefore if a high-yield ERW were feasible, its high output of nuclear radiation would be wasted.

From what has been said we see that the use of an ERW in place of a fission weapon would confer a considerable tactical advantage on one's own troops. The large reduction in collateral damage to buildings, bridges and forests means that the troops' mobility would be much less inhibited – a most important factor in post-strike operations.

Summary

The contents of this chapter are conveniently summarized in Table 3.3, which shows how the height of burst affects the principal weapon outputs.

TABLE 3.3
EFFECT OF BURST HEIGHT ON WEAPON OUTPUTS

Weapon output	Exo- > 35 km	Low air > 100 m	Surface	Subsurface
thermal	light flash only	much	less	little or none
air blast	none	much	less	less or none
ground Shock	none	slight	much	most
immediate nuclear	none	much	much	less
residual nuclear	none	some (NIA)	much (fallout and NIA)	much if explosion vents
EMP	extensive	local	local	little or none

It has been convenient in this chapter to treat the weapon outputs in the order: thermal radiation, blast, nuclear radiation, and EMP. They do not, in fact, arrive at a target in that order, and there is much overlap. Figure 3.4 shows the time history of each of the outputs received at a target 1 km away from a 27-kT low air burst. A logarithmic time scale has to be used to handle the widely different velocities and durations of the several outputs. Its origin is the instant when the first outputs (gamma rays and EMP) reach the target.

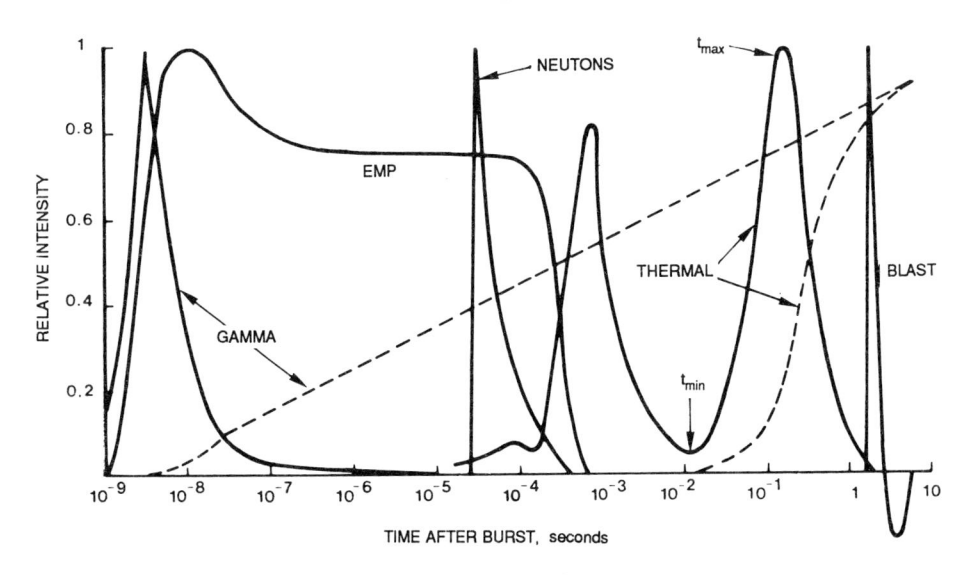

FIG. 3.4 Time histories of weapon outputs; 1 km from 27-kT burst
Dashed curves – fractions of Total gamma dose and total thermal energy

4.
Thermal Radiation and Effects

Introduction

In the last chapter the origin and nature of the thermal radiation were described. We now show how to estimate the total thermal energy at any range from a nuclear weapon, and look at the effects it may have on man and materials. The final sections of this chapter discuss the steps which can be taken to protect soldiers and their equipment against these effects.

Estimation of Thermal Parameters

The thermal threat from a nuclear explosion is specified by the total thermal energy – ultraviolet, visible and infrared radiation – which strikes a unit area at the point of interest. It is denoted by Q, and expressed in joules per square metre (J/m^2) or calories per square centimetre. We saw in Chapter 2 that the total energy released, in all forms, by the explosion of a W kT weapon is $4.186 \times 10^{12} W$ joules. It has also been mentioned that 35 per cent of the energy from an air burst appears as thermal radiation. To find Q at a range R metres from a burst we first allow for inverse square law spreading by dividing the total energy emitted as heat by the surface area of a sphere of radius R (which is $4\pi R^2$), giving us:

$$Q = \frac{0.35 \times 4.186 \times 10^{12} W}{4\pi R^2} \text{ joules/m}^2$$

This relation gives us Q only if the atmosphere is perfectly transparent to all thermal radiations, which in the real world it never is.

Absorption and Scattering

In addition to air molecules, the atmosphere contains liquid and solid particles of various kinds. The molecules of air do not have much effect over the sorts of range we are interested in, but other constituents, such as water droplets, dust particles and smoke, can both scatter and absorb the thermal

Thermal Radiation and Effects

radiations. Scattering increases the amount of energy arriving at a target other than along the line of sight from the fireball. The greater the range, the greater is the proportion of scattered radiation.

Rather more important is absorption, which reduces the thermal energy received to below the value predicted on the basis of the inverse square law. We allow for it by introducing a thermal transmission factor or transmissivity, T. The realistic expression for the total thermal energy Q is then:

$$Q = \frac{0.35 \times 4.186 \times 10^{12} \ W \ T}{4\pi R^2} \text{ joules/metre}^2$$

where W is the yield in kT and R the range in metres. If Q is wanted in calories/cm^2, the value in J/m^2 must be multiplied by 2.4×10^{-5}.

Strictly speaking, the range R in this equation (and in Figures 4.1 and 4.2) is the slant range, that is, the distance from the point of burst to the target. However, for low air bursts and targets on the ground the difference between the slant range and the ground range (the distance from GZ to the target) is small. Any unprotected target which 'sees' a low air burst at an angle of elevation greater than about 10 degrees above the horizon is likely to be destroyed. For this angle the slant range is only about 1.5 per cent greater than the ground range, so the use of the ground range instead of the slant range to find Q results in an overestimate of only 3 per cent.

The visibility is, of course, an indication of the transmissivity T. The way in which T varies with range for an average to good visibility of 20 km is shown in Figure 4.1. When these values are used in the relation for Q we obtain, for a 1-kT yield, the values of thermal energy shown in Figure 4.2. We can use this graph to find Q for any other yield since, for a given range and visibility, Q is simply proportional to the yield W. For example, if we require Q at 1,000 metres from a 5-kT weapon, it will simply be five times the thermal energy from 1 kT at this range, which we see from the graph is 110 kJ/m^2. Therefore the energy from 5 kT is 550 kJ/m^2.

The graph may also be used to find the range at which Q has some particular value for any weapon of yield W. For example, suppose we wish to know the range from a 10-kT weapon at which Q is 260 kJ/m^2 when the visibility is 20 km. This will be the range at which 1 kT gives a thermal energy of 26 kJ/m^2, which we see from the graph is 2,000 metres.

In north-west Europe visibility tends to be less than 20 km rather than greater, so that use of Figure 4.2 will, as a rule, tend to slightly overestimate Q. However, if the visibility is very poor – say 1 km – a correction must be made. The correction factor, by which Q obtained from Figure 4.2 should be multiplied, is given in Table 4.1 for a number of ranges.

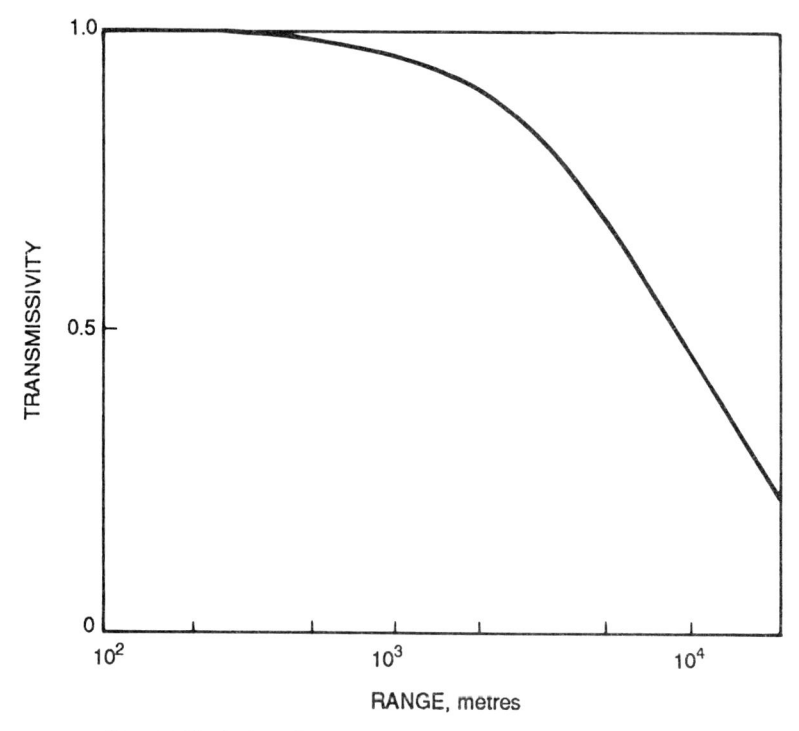

FIG. 4.1 Variation of transmissivity with range, visibility 20 km

Cloud, Snow, Dust and Smoke

We must also make corrections if the thermal radiation reaching a target is enhanced by reflection from a cloud layer above the point of burst or by fresh snow on the ground. Neither of these has much effect at the shorter ranges, less than about 1 km, at which by far the greatest contribution to Q is made by radiation coming direct from the fireball. However, at longer ranges cloud or snow reflects a great deal of thermal radiation, so enhancing Q. If either high cloud or snow is present the data from Figure 4.2, corrected if necessary for transmissivity, must be multiplied by 1.5 when the range exceeds 1 km. When both high cloud and snow are present the correction factor is $(1.5)^2$, i.e., 2.25.

There are some other circumstances which influence the amount of thermal radiation received. First, if the burst is on the surface a large amount of dust is stirred up, and the thermal energy reaching a target is reduced to half of the value for an air burst. Secondly, smoke screens such as those produced by artillery or mortar fire can absorb a considerable fraction of the thermal energy. However, this gives only coincidental protection, for if we know exactly when an enemy strike is going to occur there are more positive protective measures we can take to avoid all the weapon effects. Finally it

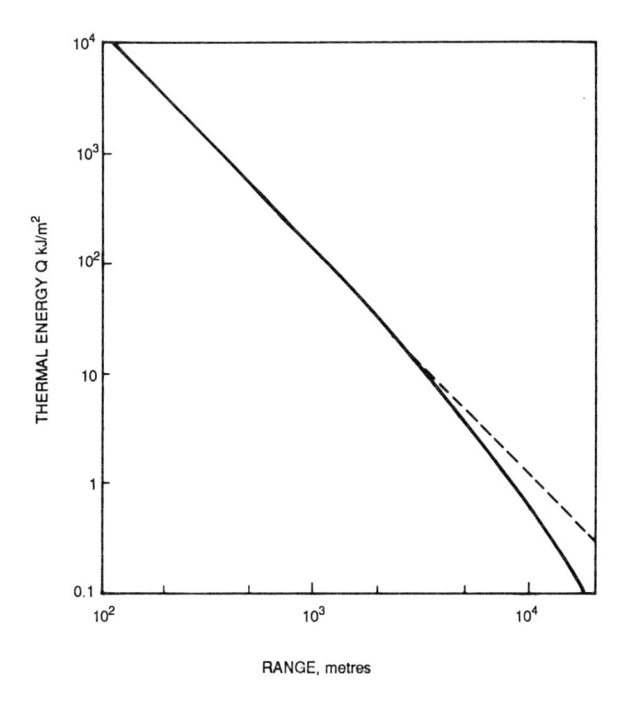

FIG. 4.2 Variation of thermal energy with range for 1 kT and 20 km visibility

TABLE 4.1
CORRECTION FACTOR FOR 1-KM VISIBILITY

Range, (km)	0.5	1	2	4	8
Correction factor	0.8	0.5	0.33	0.25	0.2

should be noted that a thin layer of mist or fog provides little protection against thermal radiation – indeed, under some circumstances it may enhance the energy received.

Other Features of Thermal Radiation

We may need more than just the value of Q if we are to be able to predict thermal effects at a given range. Information about the shape and duration of the thermal pulse may also be required. Figure 4.3 provides this for all yields of low air burst; in it time is expressed in terms of t_{max}, the time to the second maximum of observed fireball temperature, as explained in Chapter 3. The graph shows only the second pulse, which carries over 99 per cent of the total thermal energy. The time t_{max}, which increases with yield, is given by:

$$t_{max} = 0.0417\ W^{0.44} \text{ seconds}$$

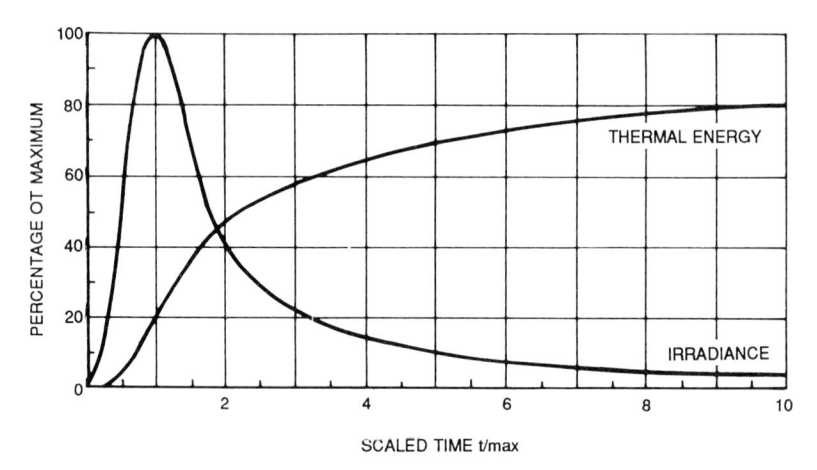

FIG. 4.3 Variation of irradiance and total thermal energy with time
(*US Department of Defense – Department of Energy*)

One curve shows how the total thermal energy builds up with time. The other shows how the rate of arrival of thermal energy (dQ/dt) varies during the pulse. This is known as the irradiance, P, in kilojoules/metre^2 sec or kW/m^2. In the figure it is specified in terms of its peak value P_{max}, which is simply related to Q (in kJ/m^2) and t_{max}:

$$P_{max} = 0.38 \; Q/t_{max} \; \text{kW/m}^2$$

The only other quantity required to complete our picture of the thermal pulse is its 'colour temperature', which is a measure of how the thermal energy is distributed over the ultraviolet, visible and infrared regions of the spectrum. For all air bursts this is about 6000 K.

Thermal Effects on Man

Effects on the Skin

It was mentioned in the last chapter that for well-equipped and well-trained men in the open nuclear radiation is the ruling or governing effect for low-yield weapons. That is to say, it is nuclear radiation which produces casualties out to the greatest range. For yields above a few tens of kT thermal radiation becomes the ruling effect, and so is responsible for the greatest proportion of casualties. The situation is different, however, if the men have a significant area of their skin unprotected against thermal radiation. For almost any weapon yield, men working in the open with their shirts off are more likely to be disabled by burns than to suffer blast or nuclear radiation injuries.

The thermal energy needed to cause second-degree burns to bare skin (which result in blistering, pain and, if they are sufficiently extensive, incapacitation) is about 160 kJ/m^2 from a 1-kT weapon – the value varies with factors such as skin pigmentation. If the yield is 1 MT the much longer pulse means that there is more time for heat absorbed by the skin to be conducted and convected away, so the energy required to produce second-degree burns rises to about 270 kJ/m^2. This, of course, does not mean that the 1 MT weapon poses a smaller thermal threat – with the aid of Figure 4.2 we can easily establish that whereas 1 kT produces a Q of 160 kJ/m^2 at about 800 metres, the 270 kJ/m^2 from 1 MT occurs at about 20 km from GZ.

Adequate protection for the skin greatly reduces the risk of thermal casualties. For a man wearing a well-designed NBC suit over combat clothing, and a respirator and gloves (as illustrated in the frontispiece), the thermal energy from tactical weapons needed to cause extensive second-degree burns is about 1.3 MJ/m^2. Thus we see that proper clothing greatly reduces the vulnerability of man to thermal radiation.

Effects on the Eye

The effects on the eye caused by the intense pulse of heat and light from a nuclear weapon are often called 'flash blindness' – a misleading term since permanent loss of sight is likely to be rare. The most important effects are dazzle and retinal burns.

The dazzle effect is similar to that experienced if one is close to a powerful photographic flash, but is much more severe. It is induced not only by light coming direct from the fireball, but also by light reflected from clouds and the ground. As a result, few people who are in the open at the time of a burst are likely to escape the effects of dazzle, no matter which way they are facing.

The consequences of widespread dazzle on the battlefield are hard to predict. Time of day will be significant, with the blindness it induces lasting longer for a burst at night when the pupil of the eye is likely to be wide open. By day those looking in the general direction of the burst will be incapacitated for about two minutes. At night this increases to about 10 minutes; in each case the consequences are serious, particularly for aircraft pilots and vehicle drivers.

It is important to realise that troops may be dazzled at ranges of tens of kilometres even from quite low yield weapons. They may therefore be affected by their own side's tactical nuclear strikes, so it is essential that they should be given warning of these as far as possible.

Retinal burns are a more severe effect, but are likely to be much less widespread. They occur when the lens of the eye focuses the fireball on to the retina. Initially they are painless, but inflammation and irritation may develop later. Slight burns heal completely, but those which are more severe may cause permanent blind spots on the retina. Vision is, however, only rarely completely lost. It has been estimated that only 2 or 3 per cent of those affected by flash blindness will receive retinal burns, and very few of these will suffer permanent partial or complete blindness.

Effects on Materials

The effect of thermal radiation on materials is influenced by the fraction of the incident radiation which is absorbed, so the nature of the surface is important. Light-coloured, smooth surfaces will absorb much less than those which are dark or rough. Except for good conductors of heat, such as metals, most of the heat absorbed will be confined to the surface. Very high temperatures may be reached, leading to ignition, especially if the material is dry, thin or porous. Thicker materials, such as vehicle tyres, plastic mouldings and heavy fabrics, may suffer only charring of the surface. Should ignition occur the fire may not persist since the blast wind, which arrives a few seconds after the thermal pulse (Figure 3.4), may blow it out. One cannot depend on this, however. Some typical energies required for ignition are listed in Table 4.2.

TABLE 4.2
APPROXIMATE THERMAL ENERGIES FOR IGNITION

Material	Ignition Energy (kJ/m^2)	
	Yield 1 kT	Yield 1 MT
khaki drill fabric	210	460
dark wool flannel	250	550
combat suit	630	1260
canvas druck canopy	630	1260
dry grass and undergrowth	80–120	200–300
dry cardboard	250	550
plastics	160–250	370–550

Although plastics have many advantages, such as low cost and weight, their use in military equipment may result in vulnerability to thermal radiation. Melting or distortion of plastic straps on vehicles or personal equipment, charring or melting of insulation on cables and antenna leads, and the melting of nylon ammunition pouches and containers could all pose serious problems after a nuclear strike.

Figure 4.4 shows a radiophotoluminescence dosimeter reader (Chapter 6) before and after exposure to a simulated thermal pulse. The side of its plastic cover which faced the simulator was destroyed by the heat, but it provided protection to the control knobs and engravings on the face of the instrument, which was still usable after exposure.

In the case of metals, conduction of heat to the interior causes a temperature rise which may not only lead to warping and buckling, but may also reduce the strength of the metal, so making it more vulnerable to the blast wave which follows. The designer of metal structures which must survive nuclear

FIG. 4.4 RPL dosimeter reader before and after thermal irradiation (*Author*)

weapon effects may have to take these so-called 'synergistic effects' of thermal radiation into account.

Optical systems are a class of military equipment which is particularly at risk from thermal effects. They are designed to be efficient collectors of visible or infrared radiation, so the thermal pulse way cause the crazing of lenses, charring of non-reflective internal coatings, and damage to filters and photocathodes.

In addition to direct damage to individual systems, a major hazard could be widespread fires caused by the ignition of grass, scrub and woodland. As well as causing burns to personnel, these could set alight fuel dumps, camouflage nets and vehicle canopies, especially as these will have been heated and dried by direct thermal radiation. The secondary dangers to personnel, ammunition, communications equipment and weapons, if this should happen, are considerable.

Protection against Thermal Radiation

Protection of Man

The thermal pulse from a low-yield weapon is so short that a man in the open will not improve his chances much by diving for cover. However, for higher yields the pulse is long enough to make evasive actions, such as falling to the ground or rolling into a ditch or trench, well worth while. Most cover, even a vehicle canopy or a sheet of corrugated iron over a trench, will provide fair protection. Substantial, all-enveloping cover will almost completely protect against thermal radiation (and will also help to mitigate the effects of blast and nuclear radiation).

As far as clothing is concerned, it is important that there should be no areas of exposed skin – hence the need for respirators and gloves to be worn when there is a risk of nuclear strikes. Several thin layers afford much better protection than fewer thicker layers. Troops wearing clothing designed for NBC warfare (as shown in the frontispiece) should have a reasonable probability of surviving thermal energies of 800 to 1300 kJ/m^2. They must, of course, be trained to roll on the ground to extinguish burning clothing.

Protection of the eyes against dazzle and retinal burns is not easy. The natural blink reflex (about 0.15 seconds) is too slow to help much for low yields. Various schemes have been proposed or tried, such as the 'Nelson patch' covering one eye (and so leaving one good eye), McCulloch-Elder spectacles (which have the outer lateral halves of the lenses darkened), phototropic optical systems (which permanently darken when exposed to intense light), and photochromic optics (which darken during the pulse but clear afterwards). However all have limitations – they are either too slow, too complex, or else do not provide sufficient protection.

A more promising approach is an electro-optical shutter incorporating a disc of lanthanum-doped lead zirconate titanate (PLZT), which becomes birefringent when an electric field is applied to it. It is used between a pair of crossed polaroid filters and a photodiode (which senses the thermal pulse and switches off the voltage applied to the PLZT element).

A simple way of reducing the recovery time from dazzle at night is to arrange that light levels, for example in the crew compartment of an AFV or aircraft, are brighter than would strictly be needed. This reduces the pupil diameter of the crew's eyes, so cutting down the dazzling effect of any nuclear strikes.

Protection of Equipment

Forethought is vital, and careful selection and testing of materials (especially plastics) at the design stage can prevent many problems. As far as possible one should avoid exposing electrical wires, hoses and plastic components to the direct thermal pulse. Metal bins should be provided for the storage of readily ignitable materials such as canvas.

Paint finishes should be highly reflective to the wavelengths in the thermal radiation. Alternatively one can use paints which either blister ('tumescent paints'), so forming a layer of thermal insulation, or which emit a dense smoke, when sufficiently heated. Types of netting which release a cloud of smoke when heated have been developed. They can be used to protect vulnerable equipment, and perhaps men as well.

Optical devices require particularly careful attention. Sensitive components should be placed in thermal expansion mounts if they are likely to be exposed to significant conductive heating. Fast-acting mechanical or electro-optical shutters may be needed to protect the users of simple optical devices, or the equipment in the case of image intensifiers and thermal imagers.

Military Significance of Thermal Effects

There is a paradox in our assessment of the thermal threat and the significance given to it by the military. Thermal effects on man and equipment may be severe out to large ranges. However, thermal radiation is easily attenuated by light cover, so that its effects as casualty producers are not considered to be sufficiently dependable for them to be taken into account in nuclear target analysis (i.e., the prediction of the effects on the enemy of one's own nuclear strikes). Enemy casualties produced by thermal radiation are regarded as a bonus.

However, when calculating safety distances for our own troops the converse is true. For certain yields and ranges the thermal threat is considered to be the governing effect. Consequently widespread and timely warning of our own intended nuclear weapon use is required to avoid injury to friendly forces, especially at night.

Conclusion

In this chapter we have seen that the thermal radiation from a nuclear weapon presents a severe threat to men in the open, particularly those with uncovered skin. Performance impairment or casualties arising from skin burns, dazzle and retinal damage may all occur out to ranges much greater than those for casualties arising from blast or nuclear radiation.

However, good training and light cover (which would afford little or no protection against the much more pervasive nuclear radiation and blast) may greatly reduce the incidence of skin burns, leaving the effects on the eyes as a continuing problem to which there are as yet no simple solutions.

5.
Blast and Effects

Introduction

In Chapter 3 the origin of the blast from a nuclear explosion and some of its features were described. This chapter examines the blast wave in more detail, looks at how its features are influenced by factors such as weapon yield, height of burst and range, and explains how it interacts with and affects various types of target. Most space is devoted to the most likely scenario – the low air burst – but the final sections deal with underground and underwater bursts.

The general nature of a nuclear blast wave is similar to that from a chemical (HE) explosion, but it is, of course, many times more powerful. Unlike a high explosive shell or bomb, a nuclear weapon does not depend on fragmentation to produce casualties and damage (indeed its case is vaporised in the explosion). However, targets on the ground are bombarded with loose objects and debris picked up by the blast wind which cause damage and casualties additional to those produced by the blast directly. Indeed, since we are most often concerned with targets on the ground, we must take into account the way the presence of the ground influences the blast wave itself. When we come to consider blast effects on targets, we shall see that the target's shape, its orientation relative to the burst, and the local topography may all affect the way the target responds. Because of factors such as these our approach to the effects of blast on targets is largely empirical.

Elsewhere in this book SI units based on the metre, kilogram, second, etc are used exclusively. In this system the unit of pressure is the pascal (Pa), which is the name given to the Newton (the unit of force) per square metre; for the nuclear blast the kilopascal (kPa) is more appropriate. However, in many publications pressures are quoted in an imperial unit, the pound per square inch (psi). In this chapter the kilopascal is used, but in some tables pressures in psi have also been given, for the benefit of those more familiar with this unit. The following equivalents may be useful:

1 kPa = 0.145 psi
1 psi = 6.895 kPa

normal atmospheric pressure = 101.3 kPa = 14.7 psi.

Air Burst Blast Wave

Velocity

The blast wave from a nuclear or HE weapon air burst is simply a rather intense compressional wave. In other words, it is akin to a sound wave, which also involves the propagation through the atmosphere of compressions and rarefactions. When we compress air its temperature rises – this is why the barrel of a bicycle-tyre pump becomes warm when we use it. The velocity of the compressional waves increases if the compression is great enough to cause heating in the compressions.

With normal sound waves the pressure changes are very small when compared with the normal ambient pressure of the atmosphere, and the heating effect is negligible. Under these conditions the velocity of a compressional wave is independent of how loud the sound is; its value at sea level is about 330 metres per second (1,100 ft/sec, 760 mph). The picture is very different in a nuclear blast wave, in which the peak pressure may be many times greater than the normal ambient pressure of 101.3 kPa. Passage of the shock then causes significant transient heating of the air, resulting in a marked increase in its velocity. The closer one is to a burst, and the higher the yield, the greater is the overpressure, and hence the temperature rise and the shock velocity. Table 5.1 gives some values which illustrate this.

TABLE 5.1
SHOCK WAVE VELOCITY

Peak static overpressure:						
in kPa	10	20	50	100	200	500
in psi	1.45	2.90	7.25	14.5	29.0	72.5
Shock wave velocity						
in m/sec	354	367	405	462	545	777
in mph	792	821	906	1,033	1,219	1,738

To see what this means in practice, consider first a thunderclap. Its sound is not intense enough to affect its velocity much, and it takes 4.8 seconds to cover a mile. The shock from a 1-kT weapon covers its first mile in 4.0 seconds. For a 1-MT yield, however, the much more intense shock wave covers the first mile in only 1.25 seconds, and the next mile takes just over 3 seconds. Thereafter, as its pressure falls, its velocity slows towards the normal speed of sound.

Static Overpressure

The principal features of the blast were outlined in Chapter 3 where Figure 3.2 showed how the pressure (relative to the normal or ambient pressure) varied with distance at a particular instant of time. Figure 5.1 shows how the

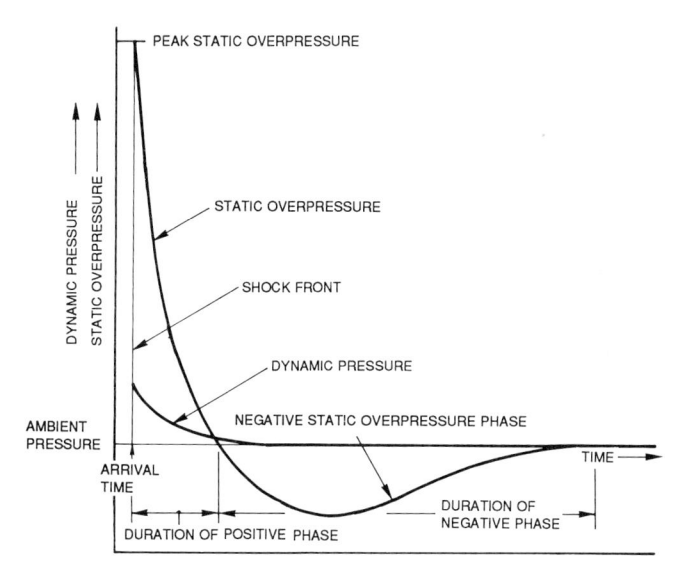

FIG. 5.1 Variation of static overpressure and dynamic pressure with time at a fixed location

pressure varies with time at a particular location. The shock front marks the boundary between the as yet undisturbed air and the region of static overpressure or enhanced pressure (the positive phase). The peak value of the pressure enhancement in the positive phase is called the peak static overpressure (PSO). Table 5.2a shows the PSO at a number of ranges from bursts yielding 1 kT and 1 MT. Note that the data in Tables 5.1, 5.2 and 5.3 all apply to bursts at the optimum height for hard targets, $60\ W^{1/3}$ metres, as defined in Chapter 3.

It hardly needs to be pointed out that nuclear explosions produce vastly greater pressures at a given range than do any conceivable weapons employing HE. For example, a 500-kg HE charge would produce a PSO of 87 kPa (12.5 psi) at 30 metres range. A 1-kT weapon yields this pressure at 300 metres; for 1 MT the range is 3,000 metres.

Positive Phase Duration

After the shock front has passed, the static overpressure steadily falls, and, after a time (called the 'positive phase duration'), the pressure momentarily passes through the ambient value. At a given range the positive phase duration increases with yield, and, for a given yield, the duration increases with range. Table 5.2b shows both these trends. The important point here is that whereas HE weapons produce (at ranges where the blast is intense enough to cause damage) a positive phase lasting for only a few milliseconds, the positive phase duration for a nuclear weapon lies between about 150 milliseconds and several seconds. We shall see later that, for an important

Nuclear Weapons

TABLE 5.2
EFFECTS OF RANGE AND YIELD ON BLAST-WAVE PARAMETERS

Yield	Range (metres)			
	300	1000	3000	10 000
a. Peak static overpressure				
1 kT	87 kPa	11.9 kPa	2.1 kPa	0.29 kPa
	12.5 psi	1.7 psi	0.30 psi	0.043 psi
1 MT	*	908 kPa	87 kPa	11.9 kPa
		130 psi	12.5 psi	1.7 psi
b. Positive phase duration				
1 kT	0.22 sec	0.47 sec	0.75 sec	0.97 sec
1 MT	*	0.73 sec	2.2 sec	4.7 sec

* Blast wave not separated from fireball

class of targets, the damage they suffer depends on the product of the overpressure and its duration. For these targets a certain overpressure from a nuclear burst is, therefore, vastly more damaging than the same overpressure from an HE weapon.

The Negative Phase

Behind the positive phase is a region of pressure below the ambient, the negative phase; it is the region from which has come the extra air which has been compressed into the positive phase. Here the pressure difference from the ambient value is much less than in the positive phase, but it lasts for several times as long. In spite of its greater duration it is nowhere near as effective at causing damage to most targets and so may usually be disregarded.

Blast Wind and Dynamic Pressure

The propagation of the compression and rarefaction which together constitute the blast wave involves a transient air motion or wind, as shown in Figure 3.3. This intense gust of wind is away from GZ during the positive phase and back towards GZ during the negative. Like any wind it exerts its own pressure, called dynamic pressure, on a target in its path. This is shown in Figure 5.1. As we would expect, its value depends on the wind velocity. The outward blast wind is strongest immediately behind the shock front, so this is where its pressure has its maximum value ('peak dynamic pressure'). During the negative phase the wind pressure is rarely significant and so is not shown in Figure 5.1.

To help one gain a picture of the origin and nature of the two pressures –

Blast and Effects

static and dynamic – the following analogy may be helpful. Consider a box sitting on the bottom of a swiftly-flowing, deep river. The water above the box exerts hydrostatic pressure on all its faces, tending to crush it. This we may liken to static overpressure. The flow of water past the box exerts a drag or dynamic pressure on the upstream face, and perhaps a suction force on the downstream face. These dynamic or drag forces which tend to move it downstream are analogous to the dynamic pressure of the blast wave. Of course, there is a difference between our analogy and the blast wave: the former is a 'steady state' situation, in which the pressures do not change with time. In contrast, the nuclear blast wave is a transient event.

Some feel for numbers is given by Table 5.3 below, which lists the peak dynamic pressures and peak wind velocities which are associated with certain values of peak static overpressure. It may be of interest to note that the range of PSO in the Table roughly matches the vulnerabilities of military systems. About 10 kPa tends to be damaging to aircraft, 20 kPa to ships, and 50–200 kPa to land-based systems. Note that this Table is quite general – a PSO of 50 kPa is always associated with a peak wind velocity of 100 metres per second (225 mph) and a peak dynamic pressure of 1.2 kPa, regardless of the weapon yield.

TABLE 5.3
DYNAMIC PRESSURE AND WIND VELOCITY

Peak static over-pressure:						
in kPa	10	20	50	100	200	500
in psi	1.45	2.90	7.25	14.5	29.0	72.5
Peak dynamic pressure:						
in kPa	0.35	1.37	8.26	31.0	110	516
in psi	0.05	0.20	1.20	4.5	16.0	74.8
Peak wind velocity						
in m/sec	23	44	100	176	292	524
in mph	51	99	225	394	653	1,172

Effect of the Ground

We are usually concerned with low air bursts, and must then take into account the reflection which occurs when the blast wave strikes the surface (ground or water). Figure 5.2 shows the position of the shock front at various times t_1, t_2, etc. after a low air burst.

At t_1, the shock front has not yet reached the surface. By time t_2 it has reached the surface and been reflected. The dashed line is the reflected wave. As we saw earlier, the air behind the incident wave will have been heated by compression. The consequence is that the reflected wave, travelling in this heated air, has a higher velocity. It therefore catches up and eventually

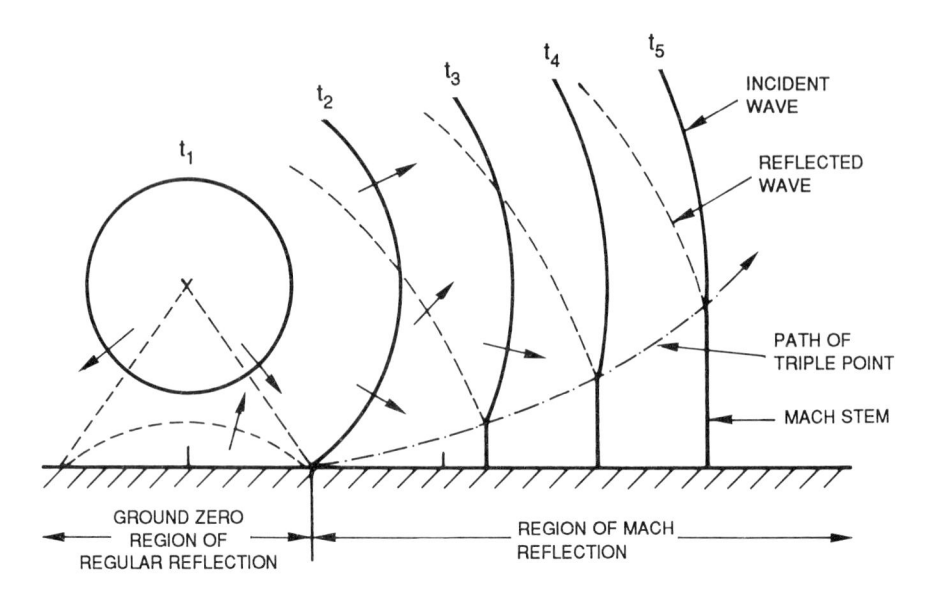

FIG. 5.2 Formation of Mach stem

merges with the incident wave to form what is known as the Mach stem. The process is called 'Mach reflection'. At time t_3 the Mach stem has extended to an appreciable height; at later times (and greater ranges) the height of the Mach stem increases further. The figure also shows the 'triple point', where the incident wave, the unmerged reflected wave and the Mach stem all meet.

This process is of more than academic interest; because of it a target on the ground experiences higher overpressures and winds than it would have if the ground has not been there to reflect the blast wave. Therefore an aircraft in flight above the region affected by Mach reflection experiences a less severe blast environment than does a tank or building at the same distance from the weapon, but which is on the ground. Instead of the single, intense shock which hits the surface target, the aircraft experiences two separate, weaker shocks, the combined effect of which is less damaging.

Blast Wave Data and Scaling Laws

As we saw in the previous chapter, thermal energies can be estimated in some circumstances by fairly simple expressions. Owing mainly to the effect of the ground this is not the case for the blast wave. Peak static overpressures on the ground produced by 1-kT weapons burst at various heights are shown in Figure 5.3.

This Figure also serves to explain in greater detail the concept of optimum height of burst. Suppose some type of vehicle requires a PSO of 105 kPa to damage it. The figure shows that a 1-kT weapon, if detonated at a height of 200 metres, produces 105 kPa out to a maximum possible range – in this case

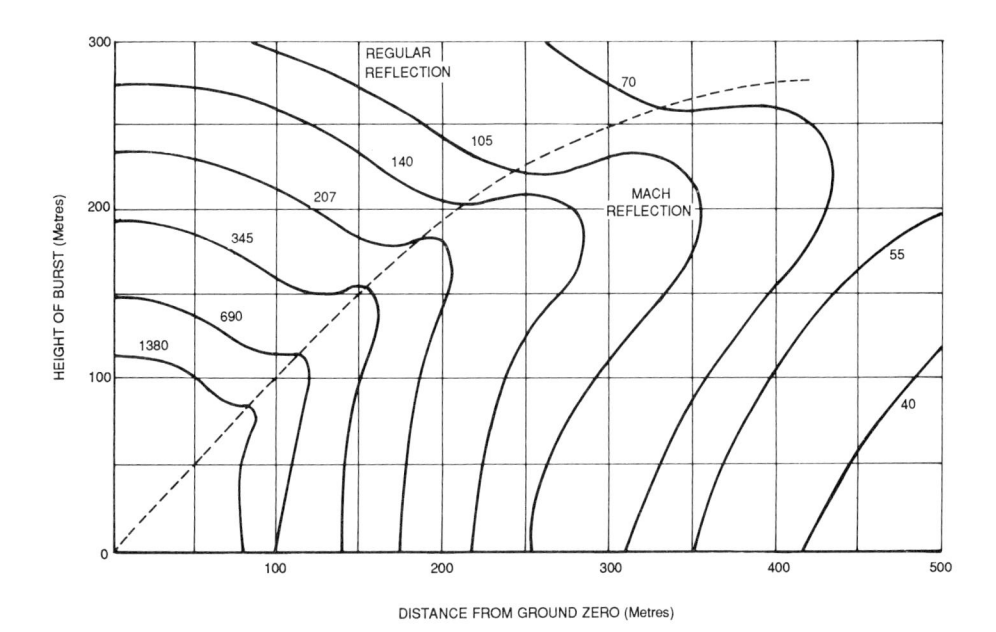

FIG. 5.3 Peak static overpressure (kPa) from a 1-kT burst

about 355 metres. A surface burst would produce the required pressure only out to 255 metres, and a burst at above 200 metres would likewise be effective out to a range less than 355 metres. So, for 1-kT yield and this particular target, 200 metres is the optimum height of burst.

Blast overpressures for other yields may be found by using some simple 'scaling laws'. The first states that if a weapon of yield W_1 (usually 1 kT) produces at a range R_1 some particular value of PSO (or some other blast wave parameter such as peak dynamic pressure, shock-wave velocity, or blast-wind peak velocity), then a weapon of yield W_2 will produce the same value of that parameter out to a range R_2 given by:

$$R_2/R_1 = (W_2/W_1)^{1/3}$$

A similar law applies to times such as those of arrival or positive phase durations at ranges where the overpressures are the same:

$$T_2/T_1 = R_2/R_1 = (W_2/W_1)^{1/3}$$

A glance at Table 5.2 will show the validity of these expressions – note that when W_2 is 1 MT, i.e., 1,000 kT, then $(W_2/W_1)^{1/3}$ is 10.0.

These scaling laws apply to mid-air bursts where the ground has no effect; they are also valid for the more usual low air burst, provided that the height of the burst is scaled in the same way. Putting H_1 and H_2 for two burst heights above ground, then these must be related by the following expression if the scaling laws for pressures and times are to be valid:

$H_2/H_1 = (W_2/W_1)^{1/3}$

Interaction with Targets

In the preceding sections we examined the features of the blast wave and how they are influenced by factors such as weapon yield, height of burst, and the ground. We now look at the way in which it affects a target. For this we consider a rectangular box (Figure 5.4) which could represent a radio set, a vehicle, a house or a ship. A shock wave approaches it from the left. In diagram (b) the shock front has just reached the left face. Two separate forces now act on this face, both tending to move the target to the right. The first is the dynamic pressure due to the blast wind, and the second, the peak static overpressure.

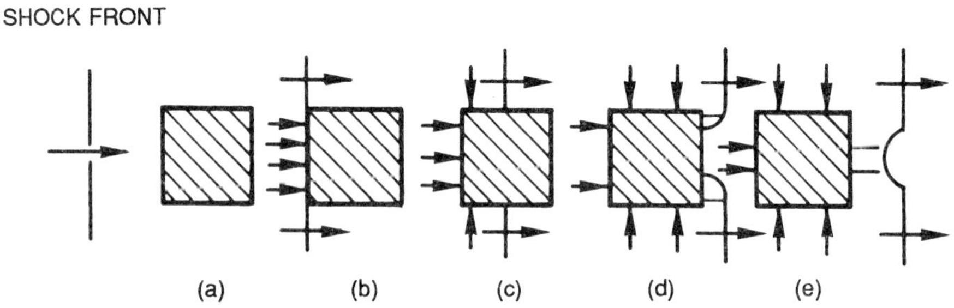

FIG. 5.4 Blast interaction with a box-shaped target

There is a complication – the blast wave is reflected at the face, and this temporarily increases the static overpressure on the left face. It is doubled for smallish pressures; for larger pressures the relative increases are greater. Table 5.4 gives the values for normal incidence (as in the diagram), or for angles of incidence within about 40° of the normal.

TABLE 5.4
EFFECT OF REFLECTION

Incident wave overpressure (kPa)	Pressure on surface (kPa)	Factor
20	44	2.2
50	120	2.4
100	275	2.75
200	660	3.3
300	1,140	3.8

At larger angles of incidence the effect steadily diminishes, and the pressure on the face falls to the ordinary peak static overpressure for a wave incident at 90°, i.e., grazing incidence, as for the upper and the lower face of the target in Figure 5.4. For a typical military target the enhanced static overpressure due to reflection lasts for only a few milliseconds, after which the static overpressure reverts to its unreflected value.

The shock front now moves along the target (diagram (c)); for a target such as a vehicle this takes a few milliseconds. At this stage the left face is still experiencing forces to the right due to both the dynamic wind pressure and the static overpressure, and the top and the bottom face are subject to crushing forces due to the static overpressure.

Thereafter the shock front spreads round behind the target (diagram (d)), a process known as diffraction, until it is completely enveloped (diagram (e)). Now the forces due to the static overpressure on the left and the right face are equal and opposite. There is no longer any displacing effect due to static overpressure, only a crushing force on all the faces. For a target such as a vehicle this stage is reached in about 10–15 milliseconds. The very short-lived displacing effect of the static overpressure is called the 'diffraction loading' of the target.

However, the dynamic pressure is still acting on the left face, and continues to do so for the duration of the positive phase, which may be several hundred milliseconds. The actual force acting on the target at any moment is the product of the dynamic pressure, the area on which it acts, and the drag coefficient of the target. This force due to wind pressure is called 'drag loading'. Thus our target has been subjected to two forces tending to move it to the right (or perhaps overturn it): they are the drag loading due to the wind acting during the whole of the positive phase, and the diffraction loading due to static overpressure, acting for the much shorter time that it takes the shock front to diffract around the target.

The response which a drag force induces in a target depends on the impulse of the force, defined as:

impulse = average force × duration

At the sort of range where buildings and equipment have some chance of surviving, the peak static overpressure is much larger than the peak dynamic pressure, but since (at any rate for small targets) it provides a displacing force for a very much shorter time, the impulse it delivers is less. Indeed, drag loading provides the dominant force tending to overturn most military targets, which are therefore said to be 'drag targets'. However, the dominant displacing force on a large building will be from diffraction loading, so it would be called a 'diffraction target'.

If our target has no apertures which allow the blast wave to penetrate (for example, if it were a box-bodied vehicle with all doors and vents closed), it will have been subjected to the crushing forces of the static overpressure for the whole of the positive phase, or until some part of it failed, allowing the external and internal pressure to equalise.

Blast Effects

Effects on Man

Except for the lowest yields, the direct effects of blast on men depend simply on the peak static overpressure. The eardrums are the most sensitive organs, and 50 per cent of exposed adults are likely to be affected by a PSO of 120 kPa (17.5 psi). The young are rather more resistant. Some lung damage is caused by 140 kPa, and 50 per cent lethality is likely to result from exposure to 400 kPa (60 psi). We shall see that these figures make man more resistant to the direct effects than some of his artefacts.

A different picture emerges when we look at the indirect effects. A man lying on the ground in the open may literally be blown away by the blast wind, a phenomenon sometimes called 'translation of prone personnel'. This is not in itself necessarily harmful: injuries occur when he strikes a tree, a building or the ground. One may expect 50 per cent casualties from this cause at a range of from 200 to 300 metres from a 1-kT weapon. For drag loading effects such as this, the range for a given effect varies as $W^{0.4}$.

One cannot ignore other indirect effects, particularly the debris carried by the blast wind, such as tree branches, glass fragments, roof tiles or bricks, which could cause injuries to men and damage to equipment. The range out to which this is important is obviously highly variable, depending on the immediate environment. A man who is inside a covered trench, tank or infantry armoured vehicle is protected from the blast wind, but he is still vulnerable to eardrum and lung damage unless the vehicle is effectively sealed, with all hatches closed and with shutters on air intakes to prevent the overpressure from affecting the interior. Within such a vehicle he is most at risk from the drag loading on the vehicle, for if it experiences a severe lateral movement when the blast strikes it, or if it overturns, then the risk of injury is high. Men in the open can be protected only by being trained to fall to the ground on first sensing the thermal flash. They are much safer inside a trench (preferably with overhead cover to keep out flying debris) or an armoured vehicle. Walls and buildings are best avoided because of the risk of collapse.

Effects on Vehicles and Equipment

Individual military targets in the field tend to be small, and may be expected to suffer from drag loading. However, some targets which are weak against compression, such as empty storage tanks, may be more at risk from the crushing effects of overpressure. With a vehicle it is often the case that external fixtures such as hatches, storage bins and antennae are damaged at ranges where the vehicle itself survives.

FIG. 5.5 Blast damage to vehicles (*Lft Col A.P. Farquar*)

Figure 5.5 shows two vehicles which were exposed to a 10-kT burst at a range of 370 m, where the PSO was 230 kPa and the peak blast wind velocity was 330 m/sec (750 mph). The tank, oriented side-on to the blast, was displaced 2.5 m with a peak acceleration of 30 g. It suffered moderate damage, principally to external fittings such as track guards and stowage bins, and its monotrailer was destroyed.

After the burst the tank was able to be driven off, and its gun was fired after sand and debris had been removed from the barrel. The lighter scout car was beyond repair. Had crews been in the vehicles they would have received a radiation dose of around 10^5 cGy. We shall see in the next chapter that they would have been incapacitated virtually instantaneously.

Some approximate figures for typical targets are given in Table 5.5. They are for bursts at optimum height, and the ranges are those at which 50 per cent of randomly-oriented targets (and orientation is obviously significant) suffer moderate damage. That is, they would require workshop repair before further use.

TABLE 5.5
RANGES FOR BLAST DAMAGE

Target	1 kT Range (m)	1 kT PSO (kPa)	1 MT Range	1 MT PSO (kPa)
tanks	170	275	2,700	150
field artillery	200	200	3,200	120
soft vehicles	300	125	4,800	60
man (prone)	240	160	3,800	100
brick houses	750	27	7,500	27
industrial buildings	350	110	5,500	50
overhead telephone lines	485	55	7,900	30
natural conifer forests	760	35	12,000	20

The ranges in the Table take no account of the damaging effect of stones and debris on vulnerable components such as vehicle radiators. Note that damage is produced by a much lower PSO (and hence dynamic pressure) from a 1-MT burst than from 1 kT, because of the much greater duration of the blast from 1 MT.

Protection of Vehicles against Blast

Because of the constraints under which he must operate, there is not a great deal that the designer can do to prevent blast from overturning a vehicle. Nevertheless, he can do much to minimise blast effects. He should ensure that the mounting pads for the engine and transmission case provide proper support even if the vehicle were moved sideways or overturned.

All caps for batteries and fuel and oil systems should be leakproof even if the vehicle were overturned. Hatch covers must have strong latches which will resist the underpressure of the negative phase of the blast wave as well as the positive phase. Antennae should be designed so that they will not break off during blast loading, or else designed so that they break at a predetermined place so that they may be easily replaced. Steps must be taken where possible to prevent missile damage to radiators and the sand blasting of optics. Inside the vehicle safety belts will reduce crew injuries, as will interior padding and the avoidance of projections. As a final point, crews must be trained to position vehicles so that advantage is taken of any protection against blast which the local topography might afford. Tethering to the ground is worth considering, particularly for high-sided vehicles.

Effects on Buildings and Forests

Buildings vary so much in design that it is impossible to generalise about their response to blast. For example, a brick-built house such as those common in Europe fails due to static overpressure effects. In contrast, modern steel-framed, light industrial buildings are affected more by dynamic pressure, and so are drag targets. Table 5.5 has damage ranges for both types.

Forests are affected by the blast wind, but they too show wide variations in response. Plantations in which the trees are of uniform height survive better than natural forests, and deciduous trees in leaf are more vulnerable than conifers. At the range quoted in the Table about 30 per cent of the trees would be blown down and extensive clearing would be necessary before vehicles could move through the area.

Effects on Ships and Aircraft

Elements of a ship's superstructure, such as radar antennae and masts, will be damaged by the blast wind. For the hull, however, static overpressure effects are dominant, and may cause shock damage to engines, propellor shafts and other machinery. Overpressure may also damage bulkheads and equipment mounted on them. At the higher levels hull rupture could occur. Typically a PSO of 25 kPa marks the onset of significant damage; hull rupture may require 75 kPa.

There are limits to what can be done to harden ships against blast since the strengthening of structures often involves weight penalties, making existing top-weight problems more severe.

Aircraft, as might be expected, are very vulnerable to air blast, particularly if it approaches from the side or the rear. Most effects are due to static overpressure, with significant damage at 6 to 12 kPa and destruction or damage beyond repair at 30 to 70 kPa. Damage results in the dishing of the skin, with, at the higher pressures, the buckling or rupture of the skin, frames and bulkheads. At these higher pressures drag forces tend to rotate or overturn aircraft on the ground unless they are securely tied down or are in hardened shelters.

Aircraft are designed and constructed to withstand a certain gust loading in normal use. This determines their hardness to nuclear blast, and there is not a great deal which can be done to improve this without incurring severe weight and performance penalties which are usually unacceptable.

Surface and Subsurface Bursts

Ground and Underground Bursts

When very hard targets such as an underground headquarters are to be engaged, ground bursts may be employed. Near GZ the blast wave from such a burst will always be more intense than it is from an air burst, but it falls off more quickly with range. The ground burst also gives rise to a shock wave in the ground, but this is not considered to be of military significance.

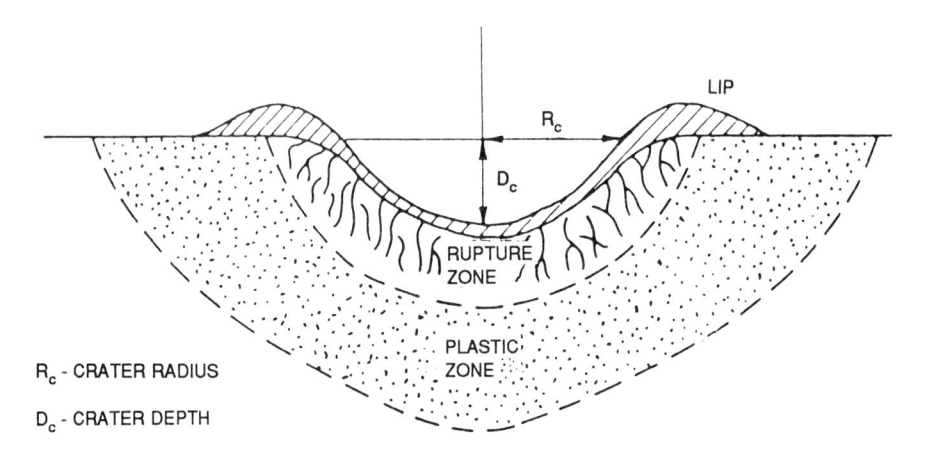

FIG. 5.6 Typical crater from an impact burst

When a ground or shallow underground burst occurs, a large amount of soil and rock is ejected, leaving a large crater (Figure 5.6). In the rupture zone there are numerous radial cracks; outside this is the plastic zone in which the rock has been permanently compressed. As with any crater formed by an explosion, there is a lip formed by debris falling back to the ground. Since highly radioactive fission products condense on the debris, the crater itself, the lip, and the area round it are a region of intense residual nuclear radiation.

As the depth of burst increases the crater dimensions increase up to some maximum, then diminish as the burst goes deeper. Table 5.6 gives crater data for surface bursts in soft, wet rock. The dimensions would maximise at roughly double those quoted if the 1-kT burst were 30 to 140 metres below the surface, and the 1-MT burst were about 250 metres down.

TABLE 5.6
CRATER DIMENSIONS FOR SURFACE BURSTS

	Yield	Crater Dimension (m)
	1 kT	1 MT
radius R_c	25	200
depth D_c	10	75
lip width	29	230
lip height	2.5	19

Underwater Bursts

The underwater shock wave from a subsurface burst differs in several important respects from the shock wave in air from an atmospheric burst. First, the sound velocity in water (1,500 metres/second) is higher than it is in air (330 metres/second). Secondly, the duration of the overpressure is rather less, although, as in air, it increases as the range and yield increase. Perhaps the most important difference is in the pressures produced. The peak overpressure at 1 km from a 1-kT burst in deep water is about 3,100 kPa (450 psi). In air the same yield at the same range produces a peak static overpressure of only 20 kPa (3 psi). The peak overpressure is given by a fairly simple relation:

$$\text{peak overpressure} = 10.7 \times 10^6 \ W^{0.37} \ R^{-1.18} \ \text{kPa}$$

where W is the yield in kT and R the range in metres.

Another significant difference in behaviour occurs when an underwater shock wave is reflected at the surface. A compressional wave in air is reflected at a denser medium, such as the ground, as a compression. As we saw in our discussion of the low air burst, the two waves reinforce, resulting in an enhancement of the blast wave 'seen' by a target on or near the ground. However when a shock wave in a dense medium, such as water, meets the less dense air at the surface, the incident compression is reflected as a rarefaction. The reflected wave reaches an underwater target a little after the direct wave and tends to cancel it – a phenomenon called 'surface cut-off', shown in Figure 5.7. Close to the surface the cancellation is more nearly complete than it is for a target at mid-depths, so the hull of a surface ship

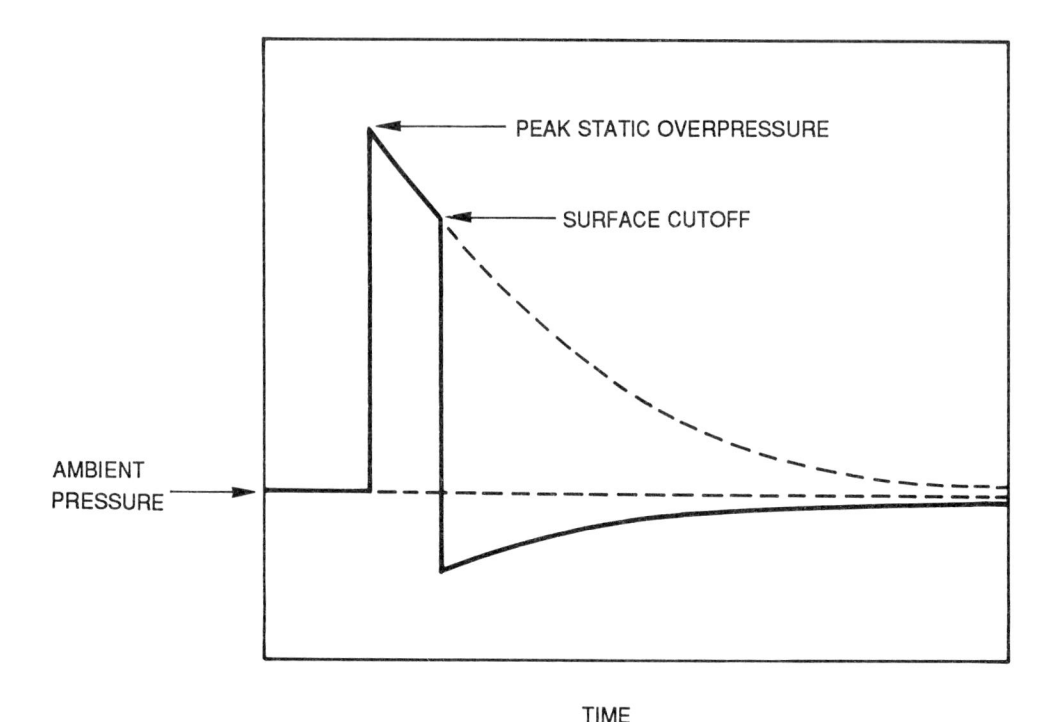

FIG. 5.7 Surface cut-off in an underwater burst

experiences a much smaller impulse from the shock wave than does a deeper target, such as a mine or a submarine.

The velocity of sound (and shock waves) in water is influenced by water temperature, and to a smaller extent by salinity and hydrostatic pressure. In the oceans the water is often in layers with different temperatures. Under these conditions sound waves are subject to considerable refraction (bending) as they move from layer to layer. This effect is important in sonar, since it means that sound energy may not propagate efficiently, if at all, into some regions. It is equally significant for underwater shock waves, so some regions experience enhanced effects, while in others shock effects are markedly reduced. It should be mentioned that analogous effects occur for shock waves in air, but they are usually too small to be important. For the underwater burst, a further complication arises in shallow water from the reflection of the shock wave by the bottom.

Damage to ships from underwater nuclear bursts is not unlike that produced by HE. The principal difference is that whereas HE tends to cause localised damage, a nuclear burst affects the whole vessel. There is distortion of the hull plating (perhaps leading to leaks or rupture) and of the frames. Hull motion under the shock puts great stress on engine and other equipment mountings, which may fail unless they have been designed to withstand these stresses. Many other components, including propellor shafts and steam or fuel lines, and weapon systems are all at risk. Damage seems to be related to the energy delivered to the hull before surface cut-off.

Conclusion

Blast – the subject of this chapter – is the weapon effect which, at least for bursts on or above the surface, produces the most immediately obvious and widespread damage to most types of civil and military installation and equipment. Indeed, in target analysis it is blast which the planners depend on to produce the greatest area of damage to material targets and (except for the lower yields) numbers of casualties.

6.
Nuclear Radiation and Effects

Introduction

In Chapter 3 the origin and the nature of the immediate and residual radiations from nuclear-weapon bursts were outlined. This chapter contains a much fuller account of these radiations, and covers the effects they have on man and on equipment. It also looks at the ways in which protection may be given against these effects by shielding and other means.

Nuclear radiation differs from the thermal and blast outputs in one most important respect. Blast and heat produce effects on man which are immediately apparent, for example, burns and broken limbs. Most of their effects on equipment are equally obvious – overturned vehicles, charred instrument panels, punctured fuel tanks and so on. In contrast, nuclear radiation effects on man may not manifest themselves until quite some time after the burst. Therefore we need instrumentation to reveal the amount of radiation which individuals have received so that commanders and medical units can estimate the future fitness of their troops and make their plans accordingly. The types of instrument used for radiation measurements in the field are covered in this chapter.

Most of the topics mentioned above cannot be examined in any detail without some understanding of the properties of the nuclear radiations associated with a nuclear explosion. Therefore we begin with a discussion of these.

Before we go any further it is as well to note that for soldiers on the battlefield, whether in the open, in slit trenches, or inside armoured vehicles, nuclear radiation is the ruling effect associated with most tactical weapons.

Nuclear Radiation Properties

The origins of the types of radiation which concern us – alpha particles, beta particles, gamma rays and neutrons – were described in Chapter 3. Unfortunately, there are fundamental differences in their behaviour, so we must discuss each separately.

Alpha Radiation

The alpha particle consists of two protons and two neutrons and is identical to the nucleus of a helium atom. These particles are emitted in the radioactive decay of a number of unstable heavy nuclei, including both of the fissile materials uranium-235 and plutonium-239. As they move through any material (solid, liquid or gas), alpha particles lose energy by ejecting electrons from atoms ('ionization') or raising electrons to higher orbits ('excitation'). Because of their large mass compared with the beta particles we consider next, their velocity is comparatively low – about one-thirtieth of the velocity of light. A consequence is that they lose energy quickly, and by the time they have traversed perhaps 40 microns in a material such as human tissue or a few centimetres of air they have been brought to rest. This means that an alpha particle-emitting material is not a hazard to man provided it is outside his body, since the particles are unable to penetrate the layer of dead cells on the surface of the skin. As we shall see though, there may be long-term consequences if the material is breathed in or swallowed.

Beta Radiation

Beta particles are fast electrons ejected in the decay of many radioactive nuclei, including all fission products and most nuclei which have captured neutrons. This 'neutron-induced activity' was mentioned in Chapter 3. Because they are very light (around 10^{-4} of the mass of an alpha particle) they travel with a velocity close to that of light, and so lose energy more slowly than do alpha particles. The consequences of their energy loss are the same, however: ionization and excitation are produced in a material they traverse. Typical ranges are up to a few millimetres in solids, or a few metres in air. Beta emitters are therefore a hazard to man whether they are inside or outside the body, though if they are outside it only the skin will be affected.

Gamma Radiation

Gamma rays are an important constituent of both the immediate and the residual radiation from an explosion. The principal source is the fission-product activity: gamma rays are usually emitted along with beta particles.

The alpha and beta particles we have just been discussing lose their energy gradually in 100,000 or more interactions with atoms they encounter. The photons of electromagnetic radiation which comprise the gamma rays behave in a fundamentally different way. They lose much or all of their energy in a single interaction with an atom, most often by a mechanism called 'Compton scattering'. Much of the photon's energy is transferred to a single electron which is ejected from the atom at high velocity. The ejected electron is indistinguishable from a beta particle, and it travels a few millimetres in solids or a few metres in air producing much ionization and excitation.

The important feature of gamma rays, however, is that many of the photons penetrate quite a large thickness of material before they lose their energy to an atomic electron. For example, we find that if the gamma rays are those present in the immediate nuclear radiation from a weapon, only half the photons are absorbed in traversing 2.9 centimetres of steel. Of the half remaining, only half (i.e., one quarter of the original number) will get through the next 2.9 centimetres of steel, and so on. The thickness which reduces the intensity of a gamma ray beam by 50 per cent is called the 'half-thickness' of the material for the particular energy of gamma rays. So there is no concept of maximum range for gammas, as we have for alpha and beta particles, and no finite thickness of absorber will stop all the gamma photons which fall on it. If we put I_0 for the intensity of gamma rays falling on a material, I_x for the intensity after passing through a thickness x, and $x_{1/2}$ for the appropriate half-thickness, then the following relation is fairly closely followed:

$$\frac{I_x}{I_0} = \frac{1}{2^{x/x_0}}$$

This applies not only for intensities (i.e., the number of photons falling on unit area per second), but also for the radiation doses and dose rates we shall shortly be discussing.

The half-thickness of materials for a particular gamma ray energy is inversely proportional to their density. Since the density of steel is about eight times the density of the body's tissues, the half-thickness for human tissue will be about 23 cm. This means that the immediate gamma rays from a weapon will affect all the body tissues, not just those close to the surface.

Neutrons

Since they have no electric charge, neutrons are unable to exert forces on atomic electrons to produce ionization and excitation in the way that the other particulate radiations (alphas and betas) do. They can only interact with nuclei direct via the three interactions described below.

The first occurs in materials consisting mainly of comparatively light atoms, such as hydrogen, carbon and oxygen; human tissue, plastics and HE are all examples. When a neutron collides with a hydrogen nucleus (which is just a proton), the proton is ejected, taking on average over half the neutron's energy. Subsequently the proton moves through the material losing energy through ionization and excitation in just the way that alpha and beta particles do. This interaction, called 'elastic scattering', is the one responsible for the slowing down of neutrons to thermal energy in the moderator of a nuclear reactor.

The ejected proton comes to rest at a point a few tens of microns from where it started, acquires a stray electron, and so becomes a hydrogen atom in a place where it is surplus to requirements. Atoms displaced in this way are, not surprisingly, known as displacements; in an organic material such as tissue they will trigger chemical changes. They are also important in pure materials such as the silicon in semiconductor devices, as we shall see.

The laws of mechanics dictate that a neutron cannot transfer much kinetic energy to a heavy atom it collides with, so elastic scatter is unimportant in materials such as steel. Instead a second process called 'inelastic scattering' occurs: a nucleus gives up energy to a heavy nucleus which then emits that energy as one or more photons of gamma radiation. These will in turn give rise to ionization and excitation, as we have seen earlier.

Thirdly, once they have been slowed down towards thermal energy (0.025 eV) neutrons are readily captured. The elements vary a great deal in their effectiveness at capturing neutrons. Some such as cadmium, boron and lithium are very efficient, while in others such as deuterium and carbon, captures are few and far between (which is one reason why these last two are good moderators). Usually the capture is accompanied by the emission of several energetic gamma rays which will produce ionization and excitation.

We have now looked at the three principal mechanisms by which neutrons interact, and have seen that all of them ultimately result in the ionization and excitation of the material. So, apart from the displacements produced when neutrons are elastically scattered in light elements, all our radiations ultimately expend their energy in the same manner.

As to how far neutrons travel – the nuclei with which they interact are small targets relative to the atom as a whole. The consequence is that neutrons travel a large and variable distance before they lose their energy and are captured. They are rather like gamma rays in this respect – there is no concept of maximum range, and a few will get through large thicknesses of most materials. Hence the neutrons of the immediate radiation will, like gamma rays, affect the whole body of an individual exposed to them.

Radiation Dose Units

The amount of damage done to human tissue or other materials by nuclear radiation depends on the number of atoms ionized or excited in each kilogram of the material. This in turn depends on the amount of energy deposited in each kilogram by the radiation. So we are led towards a unit of absorbed dose – the gray (Gy), which is the deposition of one joule of energy per kilogram of the material. When we are interested in beta particles or gamma rays it happens that a particular radiation environment will deposit very similar doses in all materials (human tissue, plastics, glass, semiconducting materials, and so on). The neutrons of the immediate radiation are another story – various materials may show marked differences in the energy deposited in them by neutrons. So when we are talking about doses from neutrons, we must specify the material involved.

The gray is an SI unit introduced fairly recently; the unit it replaces, the rad, is 0.01 joule/kilogram, equivalent to 0.01 Gy. To simplify the change, the armed forces now quote doses in centigrays (cGy), which are, of course, identical to the old unit. We shall follow this practice in this book. As well as a dose unit, we also need a unit of dose rate, that is, the amount of energy deposited in each kilogram in unit time. The unit employed is the centigray per hour (cGy/hr). Again it is identical to the former unit, the rad per hour. The total dose, dose rate and exposure time are simply related:

total dose (cGy) = dose rate (cGy/hr) × exposure time (hrs)

Radiation Effects on Man

We all know that nuclear radiation damages living tissue such as the many types of cell which go to make up the human body. The damage may be divided into two distinct categories:

a. Somatic damage, which upsets the functioning of cells, resulting in damage or death to the affected cells and perhaps, if the damage is sufficiently severe and widespread, the death of the individual. Somatic damage may itself be subdivided into early effects, which show themselves within a comparatively short time, say between a few minutes and a few weeks, and late effects, which may not become apparent until many years after irradiation.

b. Genetic damage to the germ cells produced in the reproductive organs, which causes changes, usually harmful, in subsequent generations.

When our concern is with peacetime applications of nuclear radiation, such as the industrial or medical use of X-rays, or the operation of nuclear power stations, we have to consider both the early and the late somatic effects and also the genetic effects. Radiation dose limits for workers and for the general public are set with the aim of keeping the probabilities of all these effects down to an acceptably low level. However, should nuclear weapons ever be employed in war, a commander's main concern would be how his troops are likely to perform over the following few hours, days or weeks. We shall therefore concentrate on the early somatic effects of the penetrating radiations (neutrons and gamma rays) and the effects on the skin of beta radiation.

Early Somatic Effects

Radiation alters or destroys some of the constituents of the body's cells, resulting in effects such as swelling of the nucleus, changes in the cell membranes and loss of the ability to divide (a process essential to the maintenance of life and the health of the individual). If the damage is severe enough the affected cells die.

Not all cells are equally vulnerable. Those which are most affected are the blood-forming cells in the bone marrow which maintain the supply of white blood cells. Some of these are responsible for the body's ability to fight off bacterial infection. A radiation dose of 200 to 1,000 cGy causes the white blood cell count to fall markedly, and there is a 50 per cent chance that a person receiving 450 cGy will die within a few weeks from some infection. This is known as the lethal dose to 50 per cent ('LD$_{50}$'). Blood transfusions and bone marrow transplants may improve the chances of survival.

At rather higher doses of 1,000 to 5,000 cGy the bone marrow cells are more severely affected, but the more immediate threat to the body comes from damage to the cells of the gastro-intestinal tract. The body's energy supply is upset, internal bleeding occurs, there is general circulatory collapse, and death from these and other effects occurs in a few days or so. At these doses incapacitation occurs within about an hour, and death is certain. The only treatment is palliative. In military circles 2600 cGy is assumed to cause 'immediate transient incapacitation' in 50 per cent of those receiving it.

TABLE 6.1
EARLY SOMATIC EFFECTS OF RADIATION

Dose (cGy)	Effects
Up to 150	No short-term effects.
150–250	Nausea and vomiting within 3–6 hours, lasting up to 24 hrs. Symptoms reappear 10–14 days after irradiation and last for 4 weeks.
250–350	Nausea and vomiting within 1–6 hours, lasting for 1–2 days. Symptoms reappear after 1–2 weeks and last up to 6 weeks. Up to 30 per cent deaths.
350–600	Nausea and vomiting within 1–6 hours, lasting for 1–2 days. Symptoms reappear after 1–4 weeks and last up to 8 weeks. 30 per cent to 90 per cent deaths in 2–12 weeks.
600–1000	Nausea and vomiting in ¼–½ hour, lasting 2 days. 90 per cent to 100 per cent deaths in 1–6 weeks.
1,000–2,500	Nausea and vomiting in 5–30 minutes; no latent period at the higher doses. 100 per cent deaths in 4–14 days.
5,000	Nausea and vomiting almost immediately; 100 per cent deaths in a day or two.

Above 5,000 cGy the central nervous system is the first to be affected. Tremors and convulsions occur almost immediately and death occurs within a day or two. Giving sedatives is all that one can do. These somatic effects are summarised in Table 6.1.

To provide commanders with a guide to the doses to which troops may be exposed without endangering their effectiveness, the following are the doses which may be incurred in 24 hours (current UK/US target analysis safety criteria are based on these):

negligible risk (subject to a maximum of 75 cGy in
    30 consecutive days):   5 cGy
moderate risk: 20 cGy
emergency risk: 50 cGy

It should be noted that these are wartime figures. In peacetime troops are subject to the much smaller limits which are laid down for the civil population (5 cGy per year for those who are occupationally exposed). Note also that doses received over a time of a few hours or days have strictly cumulative effects, but when doses are received over longer periods such as weeks, months or years, repair mechanisms come into effect and the somatic effects are much less than Table 6.1 suggests.

Late Somatic Effects

There are a number of late somatic effects including the induction of cataracts (opacity of the eye lens) and cancers such as leukaemia. However, these differ from the early effects in an important respect. Every individual who receives a whole body dose of 500 cGy will show symptoms of early somatic effects, and about half of those so affected will die. Of those who survive, only a very small proportion will develop late effects such as cancers perhaps 10 or 20 years after irradiation. Accurate risk figures are hard to calculate since one is looking for small increases above the natural rates of occurrence. It seems that if 10,000 persons each receive 10 cGy, or if 1,000 people each receive 100 cGy, then in each case about one or two will, at a later time, develop a cancer as a consequence of the radiation they have received. Clearly this is not of much military significance, but it is the incidence of these late effects, and of genetic effects, which set the peacetime legal limits for radiation.

Effects of Inhalation and Ingestion

So far we have been considering the effects when penetrating radiations, such as gamma rays and neutrons, coming from a source external to the body – the fireball, or fallout on the ground – irradiate all body tissues with roughly the same dose. However, there are also hazards from eating or drinking food or liquids contaminated with fallout, or breathing in air containing fallout particles. In these cases the several elements in the fallout tend to concentrate in different organs. Strontium-90 is deposited in the bones, iodine-131 in the thyroid, and unfissioned plutonium-239 in the liver and bones. The greatest risk is then of late somatic effects, such as tumours in the organs which suffer the greatest dose due to this non-uniform deposition in the body.

Genetic Effects

Irradiation of sperm cells or ova may induce mutations, i.e., changes in the information carried by the genes. These show up as new, usually undesirable, characteristics in the next and later generations. Some mutations occur spontaneously, and they can also be induced by heat or by certain chemicals such as mustard gas. The spontaneous rate is about one in 10^5 per generation; it has been estimated that if the population at large were all to receive

100 cGy the number of mutations would roughly double. It is uncertain whether the consequences for future generations would be barely detectable or disastrous.

Dose Equivalent

As far as early somatic effects are concerned, it seems that the risk of incapacitation or death depends simply on the total absorbed dose in centigrays. One may say that the LD_5 is 450 cGy without further qualification. However, matters are less simple when we look at the incidence of late effects, such as the induction of cataracts or cancers. It is found that 1 cGy of certain radiations, such as alpha particles or neutrons, is much more likely to induce these late effects than the same dose of gamma rays or beta particles.

The difference arises from the way in which the various radiations interact. Beta particles, and the electrons which gamma rays eject from atoms, both travel a distance measured in millimetres in tissue, causing light damage to several hundred cells, which may then recover. By contrast, alpha particles and the protons which neutrons knock out of hydrogen atoms travel a much shorter distance, causing much heavier damage to far fewer cells, and making late somatic and genetic effects much more likely.

In the civil world, which most concerns itself with these late effects, this dependence of damage on radiation type is allowed for by allotting each radiation a 'quality factor', which is a measure of its tendency to induce late effects, relative to gamma rays. On this basis alpha particles have a quality factor of 20, and fast neutrons 10. A new unit of dose, called the 'dose equivalent', is then introduced, measured in units called sieverts (Sv). Its value is given by:

dose equivalent (Sv) = absorbed dose (Gy) × quality factor

Legislation about exposure to radiation is primarily concerned with limiting late somatic and genetic effects, so dose limits are always expressed in sieverts. However in service circles, where it is early effects that we are concerned with, the absorbed dose in centigrays suffices.

Effects of Beta Radiation

Beta particles are emitted by fission products as they decay, and also by the radioactive species formed in soil and equipment by neutron capture ('neutron-induced activity'). They travel, at most, a few metres in air, so that those emitted as part of the immediate radiation never reach the ground. However, if the burst height is low, then beta radiation from fallout and NIA can affect exposed troops. Since the range of beta particles in tissue is at most a few millimetres, casualties are due to effects on the skin.

The worst situation is when the active material is in directly contact with the skin. If it is on the clothing the risks are less since, unlike gamma rays, beta particles are absorbed to some extent in passing through clothes to the skin. The wearing of masks and gloves is therefore essential when fallout is present. Beta particles from fallout on the ground or equipment will be further attenuated as they pass through the air so are less of a hazard, particularly if the skin is covered.

TABLE 6.2
EFFECTS OF BETA RADIATION ON SKIN

Dose (cGy)	Effect on Skin
600	No effects.
2,000	Moderate reddening and soreness.
4,000	Reddening and soreness within 24 hours. Skin breakdown in two weeks.
10,000	Severe reddening and soreness within 24 hours. Skin breakdown in one week.
30,000	Severe reddening and soreness in 4 hours. Severe skin breakdown in a few days.

The effects they produce on the skin are rather similar to the effects of heat, so they are sometimes called 'beta burns'. They can, however, be more severe than conventional burns since the whole thickness of the skin may disintegrate, leaving the nerve endings exposed. They are incapacitating, though not fatal. Table 6.2 shows how beta radiation affects the skin. Healing is usually complete, but may take several weeks or months.

The doses in the Table may suggest that beta burns are not much of a problem because of the very high doses necessary to induce them. However, typical beta-decaying elements emit roughly comparable amounts of beta particle and gamma ray energy. The beta particles deposit all their energy in a few millimetres of tissue, whereas the gamma rays deposit most of theirs in a thickness measured in hundreds of millimetres. Thus the beta particles deposit their energy in a mass of tissue which is only from a fiftieth to a hundredth of the mass which receives the gamma ray energy. Therefore the beta dose to skin from fallout upon it is about 50 to 100 times the gamma dose to the deeper tissues. This means that a beta dose of 10,000 cGy to the skin would be associated with a whole-body gamma dose of only 100–200 cGy. In this example the whole-body gamma dose would not cause very serious early somatic effects, but the beta burns to the skin could be disabling.

Immediate Radiation from Weapons

The origin of immediate nuclear radiation from weapons was discussed in Chapter 3; the components were seen to be neutrons and gamma rays. We now look at these in turn to see how the dose from each is affected by the range and yield for a low air burst weapon. It may appear to be unnecessarily pedantic to discuss the gammas and neutrons separately, since the early somatic effects on man depend only on the total absorbed dose, regardless of whether this is due to just neutrons, just gamma rays, or (as is usually the case) a mixture of both, but there are sound reasons. First, we shall see in a later section that neutrons and gamma rays produce entirely different effects in the semiconductor devices used in electronics, and, secondly, shielding materials are usually rather more effective at reducing the gamma dose than they are against neutrons. For these reasons we need separate data on neutron and gamma doses.

Neutrons

The way the neutron dose from a 1-kT weapon falls off with range is shown in Figure 6.1. In this graph the range is not the ground range (from GZ) but the slant range. However, for the reasons discussed in Chapter 4, it will usually be accurate enough to assume that ground range and slant range are the same. The scaling law is simple since the neutron dose at a given range is closely proportional to the yield, so the data may be used to provide the neutron dose from any weapon (except enhanced radiation weapons). Thus at a given range a 20-kT weapon gives a neutron dose which is 20 times that read from the graph.

For a fission weapon neutrons of thermal energy (0.025 eV) up to about 10 MeV are present at all ranges. This comes about because fast neutrons are slowed by collision as they travel out through the air, and slow neutrons are captured. These two effects are nearly in balance, so the energy spectrum changes very little with range. Typically the first neutrons (which are nearly all emitted during the chain reaction) reach a target in a little under a millisecond; after one second virtually all the neutron dose has been received.

Gamma Rays

As we have just seen, the neutrons are emitted early on, before the blast wave has developed, so they travel through undisturbed air. Some of the gamma rays have a less simple journey. Those emitted during fission and those generated by neutron interactions with the weapon materials and the atmosphere travel through undisturbed air just as the neutrons do.

However, a significant proportion of the gamma rays are emitted by the fission products in the fireball after the blast wave has developed. At this time a large amount of air has been displaced from the negative phase of the blast wave into the positive phase, where it is further away from the fireball. To understand the consequences for the gamma rays let us look at a rather extreme analogy. Suppose we have a solid lead sphere 10 cm in radius with a small hole in the middle in which is stored a gamma-emitting source. The gamma rays would have to penetrate 10 cm of lead to escape from the sphere, and only a small proportion would make it. If the same quantity of lead were refabricated into a hollow spherical shell 100 cm in radius it is simple to show that its thickness would be only 0.33 millimetres. Virtually all the gamma rays would penetrate it. Applying these ideas to the nuclear explosion, we see

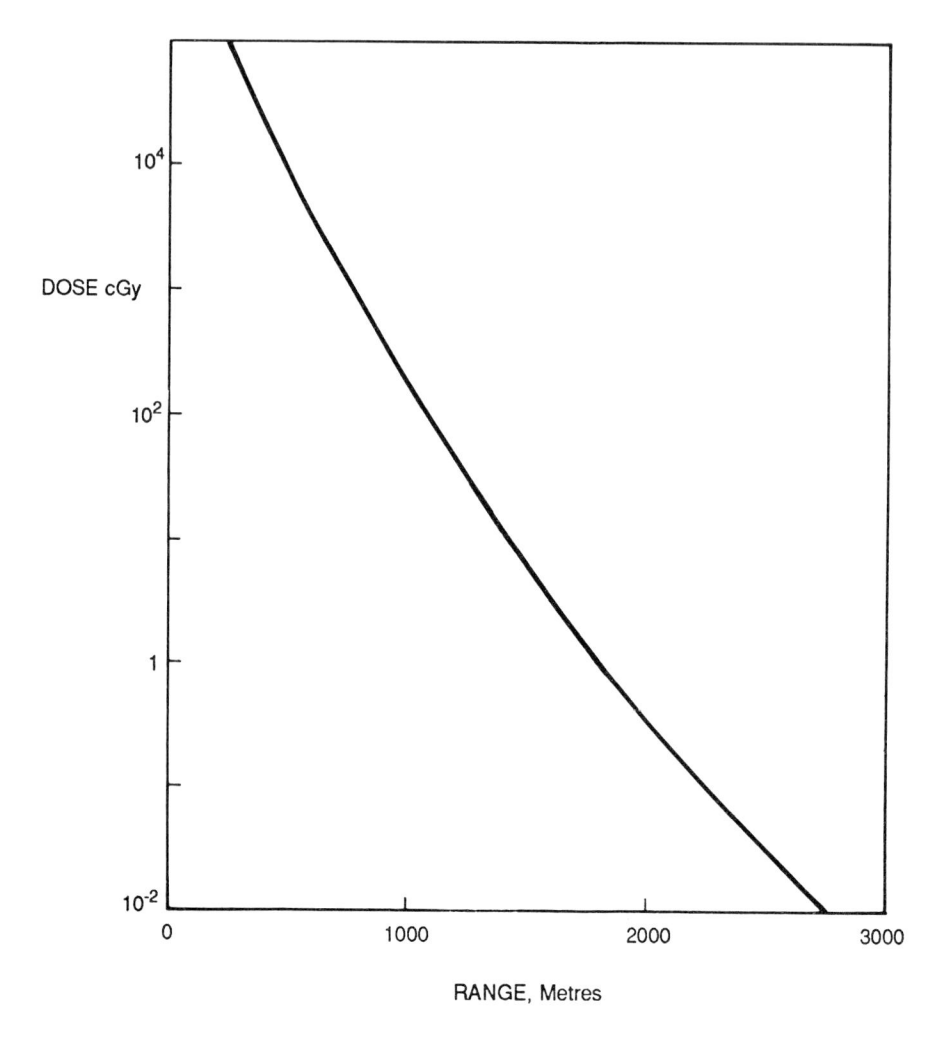

FIG. 6.1 Neutron dose from a 1-kT weapon

that the outward displacement of air associated with the blast wave results in the gamma rays being much less attenuated in their passage through the air than they would have been if the atmosphere had been undisturbed. The higher the weapon yield, the more intense is the blast wave, and the greater is this effect, known as 'hydrodynamic enhancement'.

For these reasons we cannot get by with a single graph and a simple scaling law to give us the gamma ray dose at any range from any weapon yield, as we can for neutrons. Instead, Figure 6.2 gives data for various yields between 1 kT and 1 MT. Interpolation can be used for other yields.

With the aid of Figures 6.1 and 6.2 it is easy to find the total immediate dose at a given range from a weapon. However, finding the range at which a

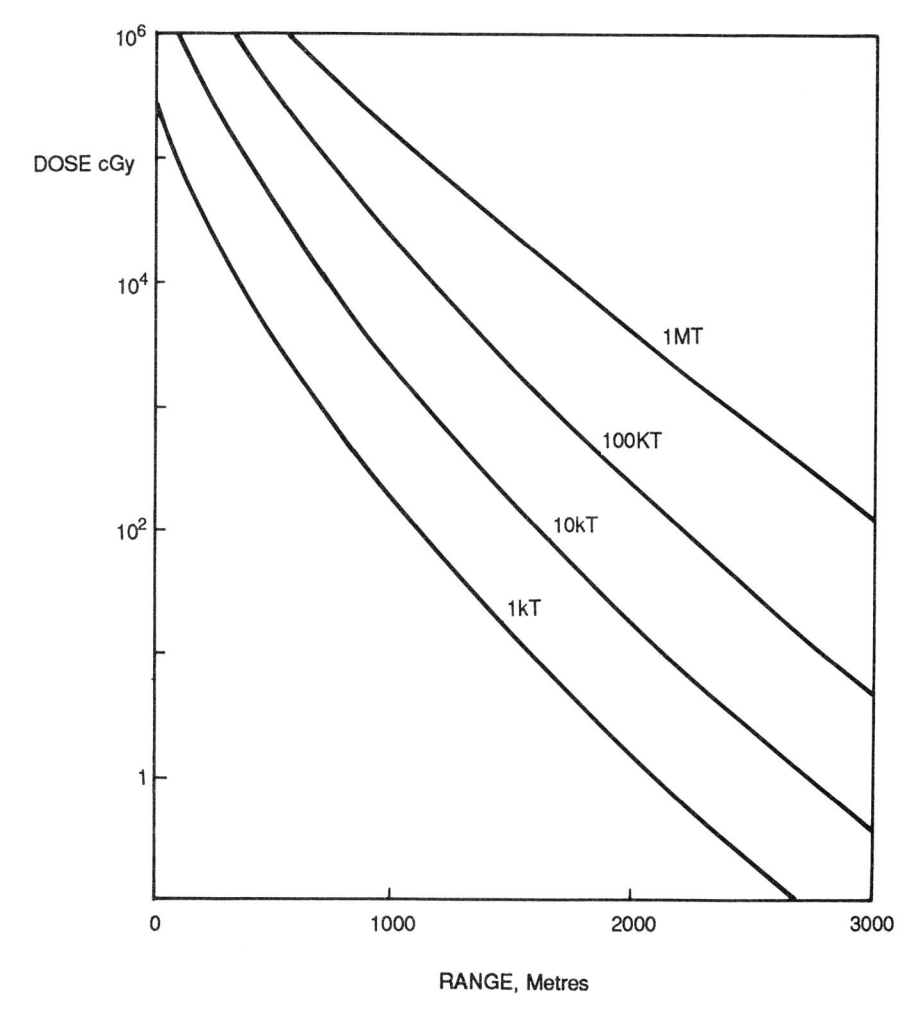

FIG. 6.2 Immediate gamma ray dose

particular weapon will produce some given total radiation dose is not so straightforward – a trial and error approach is needed.

Rate of Delivery

As we have seen, several mechanisms contribute to the immediate gamma rays. Consequently (unlike neutrons) they continue to arrive at a target over quite a long period. This is indicated in Figure 3.4 and shown in much more detail in Figure 6.3.

The prompt gamma rays are emitted during the actual fission process. They are followed by gamma rays generated by neutron interactions with the

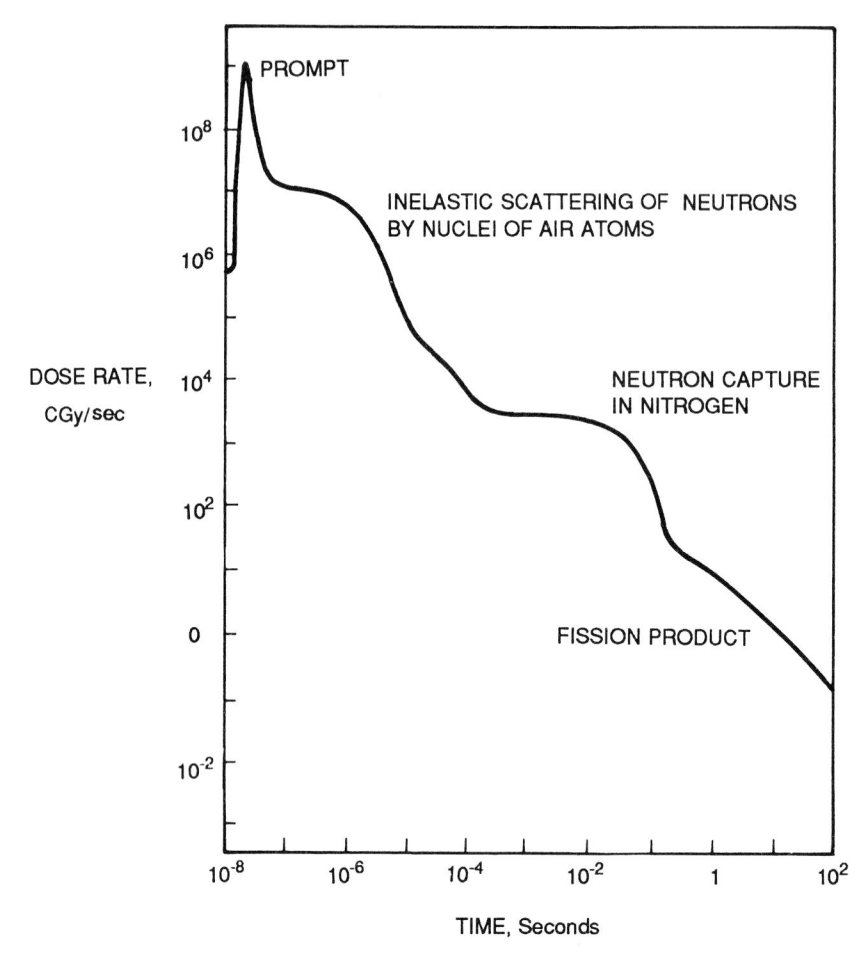

FIG. 6.3 Gamma dose rate 300 m from 1 kT
(*US Department of Defense – Department of Energy*)

atmosphere. Last, but by no means least (logarithmic scales such as the time scale in Figure 6.3 can be misleading), come the gamma rays emitted by the fission products. The intensity of this last group falls with time due to the decay of the fission products and the rapid rise of the fireball containing them. Table 6.3 shows how the immediate gamma dose accumulates with time for two weapon yields.

We see from Figure 6.3 that at 300 metres from a 1-kT weapon (where the total immediate gamma dose is about 3,000 cGy) the gamma dose rate has the extremely high initial value of about 10^9 cGy/second; this is maintained for some tens of nanoseconds. It contributes less than 1 per cent to the total immediate gamma dose, so is not significant for man. However, when we come to look at radiation effects on electronics we shall see that it may have serious consequences for semiconductor devices.

Nuclear Radiation and Effects

TABLE 6.3
RATE OF DELIVERY OF IMMEDIATE GAMMA RAY DOSE

Yield	Range (metres)	Total gamma dose (cGy)	\multicolumn{5}{c}{Dose in cGy delivered in specified time (seconds)}
			0.1	1.0	2.0	5.0	10.0
10 kT	1,000	2,000	700	1,300	1,500	1,800	1,980
5 MT	2,500	2,000	0	100	330	1,000	1,800

Effect of Weapon Type

For blast and thermal radiation one may assume that the output of each depends solely on yield, regardless of weapon type (except for the enhanced radiation weapon; see Chapter 3). This is not the case for immediate radiation. For example, a particular yield could be obtained from an implosion weapon, or from a boosted fission with a thinner tamper (which would allow more neutrons to escape). Therefore the radiation data in Figures 6.1 and 6.2 can be taken only as broadly representative. Some weapon types could give radiation outputs between 0.25 and 2.0 times those deduced from the figures.

Effect of Burst Height

Figures 6.1 and 6.2 apply to an air burst weapon in which the fireball does not touch the ground. With a surface burst the neutron and the gamma ray dose are affected differently. The neutron dose may be taken to be half that from an air burst. For yields of 50 kT or less the gamma ray dose is two-thirds of that from an air burst. For 100 kT a surface burst and an air burst yield about the same gamma dose. For yields above 300 kT the gamma radiation is enhanced by a factor which increases with yield – it is about 2.0 for 2 MT.

Residual Radiation

By definition residual radiation is that which is received later than one minute after a burst. Two sources contribute to it: they are neutron-induced activity and fallout. The former is usually a fairly minor problem, but this is not so for fallout. All surface or air bursts produce fallout somewhere at some time. However, if the burst is at or above the optimum height for hard targets ($60\ W^{1/3}$ metres) the fireball does not touch the ground, and the fallout, being world-wide and slow to settle, is not a military problem.

We shall therefore concentrate on surface bursts, very low air bursts, and shallow underground bursts, all of which result in fallout which is local, intense, and a major hindrance to subsequent operations in the affected area.

Local Fallout

When a shallow underground, surface, or very low air burst occurs a large amount of soil is drawn up into the fireball, providing condensation centres for the vaporized weapon debris. This is carried upwards in the cloud which eventually stabilizes at a height which depends on the yield. For a 100-kT yield the cloud stabilizes with its base at about 6 km above the surface and its top at 12 km. Most of the radioactive fission products are in the upper part of the cloud stem and the lower part of the cloud itself.

The fission products are a complex mixture. Over 300 radioactive isotopes of 36 different middle-weight elements have been identified in them. These produce the bulk of the activity, and one hour after a 1-kT explosion there are about 10^{19} disintegrations occurring per second, involving the emission of beta particles and gamma rays. There will also be some emission of alpha particles from unfissioned plutonium or uranium, but these are few in comparison. For example, a kilogram of plutonium-239 emits 10^{12} alpha particles per second; the longer-lived uranium-235 emits far fewer.

In Chapter 1 we saw that every radioactive species decays with time according to an exponential law, its decay being characterised by a unique half-life. The complex mix of fission products does not follow any simple theoretical law. However, measurements show that for up to six months after an explosion the dose rate from fission products is given by the following expression:

(dose rate t time units after burst) = (dose rate 1 time unit after burst) $\times t^{-1.2}$

From this we can derive a rather simpler rule, called the 'seven and ten rule' – if the time after the burst is multiplied by seven, the fission product dose rate falls to one-tenth. As an example, let us assume that the dose rate at some point due to fallout is 500 cGy/hour one hour after a burst. Then:

at 7 hours after the burst it will be 50 cGy/hr
at 49 hours (about two days) it will be 5 cGy/hr
at $7 \times 49 = 363$ hours (about 2 weeks) it will be 0.5 cGy/hr.

Another handy expression follows from this law – the total fallout dose received from one hour after a burst to an infinite time is simply five times the dose rate at one hour. In the example above it would be 2,500 cGy.

Of course, the fallout at a point downwind of a surface burst does not all arrive simultaneously, since heavy particles will fall more rapidly. In practice the dose rate at a downwind point will vary with time in the way shown in Figure 6.4. In this example the fallout begins to arrive 1.5 hours after the burst, and is complete at 3.5 hours after it. Thereafter its decay is according to 'the seven and ten law' (the straight part of the graph).

When fallout occurs a commander will wish to know its extent and intensity as soon as possible so that he can move troops from the worst-affected areas. Monitoring of dose rates from fallout is therefore essential (the instruments used are discussed later in this chapter). However, the measurements he receives will have been made at widely different times, so they cannot be plotted on a map direct to enable dose rate contours to be drawn. Instead, the

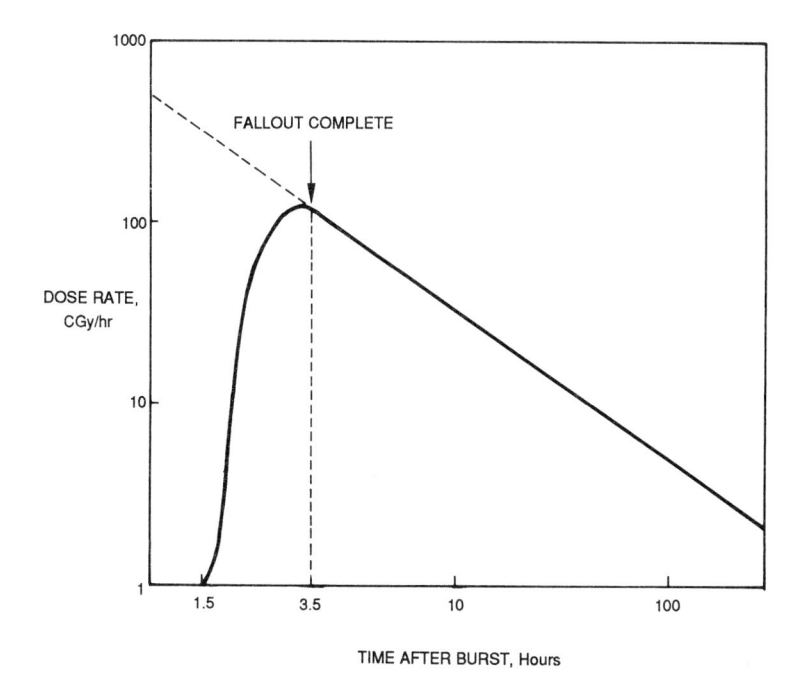

FIG. 6.4 Fallout dose rate variation with time

seven and ten law is used to find what the dose rate would have been one hour after the burst. As an example, at the position for which the dose rate is shown in Figure 6.4, any dose rate measurement made more than 3.5 hours after the burst yields a one-hour dose rate of about 520 cGy/hr; this is the figure which would be plotted. The fact that there was no fallout actually on the ground at the particular point one hour after the burst is immaterial.

Dose rates from fallout may be high enough to be life-threatening. A 1-kT surface burst could produce dose rates of 1,000 cGy/hr or more over an area extending several kilometres downwind from GZ. With high yields the threat is much more serious. For example, a 15-MT device, burst on the surface of Bikini Atoll in the Marshall Islands, produced fallout which arrived at Rongelap Atoll 200 km downwind about seven hours after the burst. At one place on Rongelap the dose rate peaked at 150 cGy/hr, and the total dose from it in the first four days was 2,000 cGy (several times the lethal dose). Even 450 km downwind from GZ 100 cGy was accumulated in this time.

Because of the threat which fallout presents, it is obviously unsatisfactory for a commander to wait until he has a complete fallout plot based on dose rate measurements before he plans the redisposition of his troops, for by that time much damage will have been done. Therefore methods of predicting the likely extent of fallout have been developed so that troops at risk can be moved (or at least warned) before the fallout actually arrives. To make a fallout prediction we must know not only the location of a burst but also its yield (since this determines the amount of radioactivity in the cloud and the height to which it rises). We also need the burst height, which determines whether the fallout will be local (and hence of military significance) or global. Instruments called 'nuclear burst indicators' have been developed. Typically they use the arrival time of the EMP at three separate points to provide the location, and the time to thermal maximum to give the yield. We also need the wind velocity and direction at all heights up to the top of the stabilised cloud.

Prediction can do no better than indicate the areas most likely to be affected because of the simplifications which are made in the prediction process, the lack of burst height data, errors in the yield or cloud height estimates, and the fact that wind information is usually a few hours old. Rain occurring while the fallout is descending may also cause anomalies, such as local hot spots. For these reasons predictions must always be confirmed by the actual monitoring of dose rates.

Neutron-induced Activity

The origin and the nature of the neutron-induced activity from a surface or low air burst have been described in Chapter 3. It is much more localised than fallout, affecting only a small, circular area around ground zero. Its extent in soil and equipment is very dependent on the concentration of certain particularly susceptible elements, such as manganese, aluminium, silicon, sodium, copper and zinc, so neither its extent nor its rate of decay are easy to predict. It must be appreciated that when a burst occurs at a height such that the fireball does not touch the ground, there will not be much local fallout, but there will still be a risk of neutron-induced activity. Generally it will not create many problems except close to GZ. Nevertheless an old GZ should be monitored for radiation before being reoccupied.

Residual Radiation Doses to Aircraft and Ships

Aircraft in flight (and their occupants) may be exposed to radiation if they fly through the radioactive cloud from a nuclear explosion. The dose received will depend on the explosion yield, how long ago it happened, and, of course, the aircraft's speed.

For ships a surface burst over water delivers an immediate radiation dose much the same as that from a burst over land. Residual radiation effects are broadly similar, but condensation nuclei, which now consist of salt particles and water droplets, tend to be smaller. This results in less close-in fallout and lower dose rates than for a burst on or over land. Serious problems are unlikely for yields below 100 kT.

Things are less simple when the burst is underwater. None of the neutrons or early immediate gamma rays will reach a surface target, but the later gammas from fission-product decay, supplemented by gammas from NIA

Nuclear Radiation and Effects

(mainly due to sodium) produce effects on the surface which do not neatly divide into immediate and residual.

The first radiation received is from fission products and NIA in the initial plume. When this collapses some of the activity is transferred to a doughnut-shaped pool of water round the GZ. This drifts with the current, steadily losing activity as it expands, mixes and decays. However, most of the activity is in a dense, fog-like aerosol, called the 'base surge', which moves outwards from GZ with the wind. Although very high dose rates exist within it initially, some 30 minutes later they are comparatively low.

Nuclear Weapon Accidents

The accidental detonation of a nuclear weapon resulting in the release of its planned yield is most unlikely because of the many safety features which these weapons contain. The greatest risk is from the crashing of an aircraft carrying a weapon. A fire is very likely, and the HE in the weapon may burn or partially explode. This causes dispersal of the fissile material over an area of perhaps a few square kilometres. There will not be much external hazard to man from this because of the short range of the alpha particles emitted by fissile materials and the comparatively few gamma rays which accompany them. However, if contaminated dust is breathed in, or contaminated food or water consumed, then the fissile materials tend to concentrate in certain organs, particularly the liver, lungs and bones, and malignant tumours may eventually result. Thorough removal of contaminated soil is essential.

Radiation Effects on Materials

The only materials which are as sensitive as man to nuclear radiation are the semiconductor devices of modern electronics, and certain optical systems, particularly long fibre-optic cables. The general effects are known as transient radiation effects on electronics ('TREE'), but it is important to note that it is the radiation which is transient – the effects may be permanent. Since as a rule gamma rays and neutrons produce quite different effects, we must treat them separately.

Immediate Gamma Ray Effects

At the beginning of this chapter we saw that gamma rays cause ionization in materials that they pass through, that is, they eject electrons from atoms, so creating free electrons and positive ions. If the material is a semiconductor, the latter are called 'holes'. Both the electrons and the holes move if a voltage is applied, so a current flows, its magnitude depending on the dose rate.

We have seen (Figure 6.3) that, when the first gamma rays from a weapon reach a target, the gamma dose rate peaks momentarily at an extremely high value (around 10^9 cGy/sec) for some nanoseconds. The consequence in a semiconductor device such as a diode with reverse bias is that a very large current, called a photocurrent, flows while the dose rate remains high. This causes not only corruption of data, but may also result in permanent effects such as the burnout of the device and damage to the power supply feeding it.

Similar effects occur in transistors and integrated circuits. With the latter there is also the possibility of SCR action or 'latch-up', that is, photocurrents which continue to flow permanently. The current only ceases when the power supply is switched off, or the power supply or the integrated circuit burns out. Even without the occurrence of latch-up there is a risk that photocurrent may cause burnout of the thin metallization interconnections in integrated circuits.

There are a number of steps which may be taken in the design of electronics to harden them against photocurrent effects. First, devices vary a good deal in their susceptibility, so those types least affected should be selected. Secondly, current-limiting resistors can be incorporated. If these steps are unable to provide the necessary degree of hardness, photocurrent effects can be circumvented by using a fast radiation detector which removes power from vulnerable circuits when the dose rate rises above a certain level.

An additional gamma-ray effect which depends on the total dose is found in some devices – a buildup of charge in dielectrics, such as the passivation oxide layer in MOS devices. To prevent this one must avoid using susceptible technologies.

Neutron Effects

Semiconductor devices consist of extremely pure silicon with a small, carefully controlled level of selected impurities. It is these imperfections in the crystal lattice which determine the characteristic of the device. We saw at the beginning of this chapter that neutrons give rise to atomic displacements. In silicon the result is a reduction in minority carrier lifetime; in bipolar transistors this results in a marked loss of gain and changes in other device parameters. The initial large parameter changes partially anneal in a few seconds, leaving smaller (but still significant) permanent effects. Information on device vulnerability is available, enabling designers to select devices which are the least affected, and to design their circuits so that they will tolerate the expected changes in the device parameters.

Effects on Glass and Other Materials

Both gamma rays and neutrons cause loss of light transmission in glass, principally at the blue end of the spectrum, so that the glass takes on a brownish colour. Transmission in the infrared is much less affected. Glass composition strongly influences the loss, which depends on the total dose. The transmission tends to recover with time, most rapidly when the temperature is high, but there may be a residual permanent effect.

As a rule this is not too serious, except in vision or other devices employing a large thickness of glass. Just such a system is a fibre-optic cable in which the optical path in glass may be many kilometres long. However, by careful fibre selection and testing, it is usually possible to produce systems which will survive, though transmission is likely to be lost during the gamma-ray pulse due to photocurrents induced in the detector at the end of the cable.

The only other materials which show a susceptibility to radiation anywhere near those of living tissue, semiconductors and glass are certain plastics, which may suffer changes in their mechanical and electrical properties. None are likely to be much affected by the doses we are concerned with.

Residual Radiation Effects

Since residual radiation contains no neutrons, and since gamma dose rates, at perhaps 1,000 cGy/hour, are many orders of magnitude below the 10^9 cGy/second which can be reached in the immediate radiation, there are no problems from displacements and photocurrents in semiconductor devices. Nevertheless fibre-optic cables are still at risk, particularly if they are surface-laid and long, and additionally the performance of some night-vision devices such as image intensifiers is severely degraded by gamma dose rates similar to those associated with residual radiation.

Radiation Instrumentation

Since the effects which radiation produces on man do not, unlike the effects of heat and blast, show themselves immediately, there is a need for dosimeters (devices indicating the total radiation dose), the readings from which enable commanders to assess the likely future fitness of troops to undertake tasks. It is not sufficient to provide a platoon with a single dosimeter because individual soldiers could receive doses varying by a factor as high as 100, depending on their closeness to the burst and whether or not they were shielded, by, for example, being in a trench, dugout or armoured vehicle at the time of a nuclear strike. Individual personal dosimeters are needed, to record the total dose from both immediate and residual radiation. Units also need dose rate meters to give them prompt warning of the presence and intensity of residual radiation.

Personal Dosimeters

Many radiation detection and measurement systems make use of the ionization produced in gases when they are irradiated with gamma rays. If a voltage is applied between a pair of electrodes in the gas, the ions move, constituting a current through the normally insulating gas. This arrangement is called an ionization chamber; it can be designed so that the current is proportional to the radiation dose rate. The quartz fibre dosimeter, which is a miniature version of a gold-leaf electroscope, works on this principle. A charged fibre is viewed through lenses against a calibrated scale. Since it measures charge rather than current the reading is proportional to the total dose. It is alone among the dosimeters to be discussed in being 'direct reading' – the wearer can take its reading at any time without needing any other equipment. However, it suffers from several limitations, the most serious being that it does not respond to neutrons.

Photographic film badges, in which the blackening of the film is proportional to the radiation dose, and the phenomenon of thermoluminescence (the emission of light by certain materials when they are heated after having been irradiated) are both widely used in the civil field for personal dosimetry, but both have characteristics which make them somewhat inappropriate for service use. For example, photographic films must be developed under carefully controlled conditions – a process which takes some time – and the dose stored in a thermoluminescent dosimeter ('TLD') is destroyed in the act of reading it.

Perhaps the most suitable dosimeter for military use is the radiophotoluminescence ('RPL') dosimeter. It consists of a block of a special type of phosphate glass which, when it is illuminated with ultraviolet light, emits visible light to an extent which depends on the total gamma dose it has absorbed. Like some other types, it does not respond to neutrons, but this limitation is overcome by incorporating with it a semiconductor diode the electrical characteristics of which are affected by the neutron dose. It is, of course, indirect reading – the reader measures both the fluorescence of the glass and the electrical characteristics of the diode. Neither the glass nor the diode is affected by the reading, so an individual's dosimeter carries a permanent record of the radiation dose it has received. An RPL reader and dosimeters are shown in Figure 4.4.

Dose Rate Meters

Dose rate meters (sometimes known as radiation monitors or survey meters) are designed to measure the gamma dose rate from residual radiation. Usually the small current through an ionization chamber is amplified to provide a measure of the dose rate. These meters take a number of forms, depending on how they are intended to be used. The most common is a pocket-sized portable unit; a typical example is shown in Figure 6.5. If it is to be used for a radiation survey from a vehicle, the operator must stop and move well away from it before taking each measurement, since the vehicle is likely to pick up contamination, particularly under the wheel arches, which would render meaningless any measurements taken from the vehicle.

To speed the survey process ionization chambers are mounted on reconnaissance vehicles with a heavy shield under the chamber to intercept gamma rays coming from any contamination on the vehicle. This ensures that the chamber responds only to contamination on the ground around the vehicle.

Rapid surveying after any surface burst is obviously vital, so dose rate meters

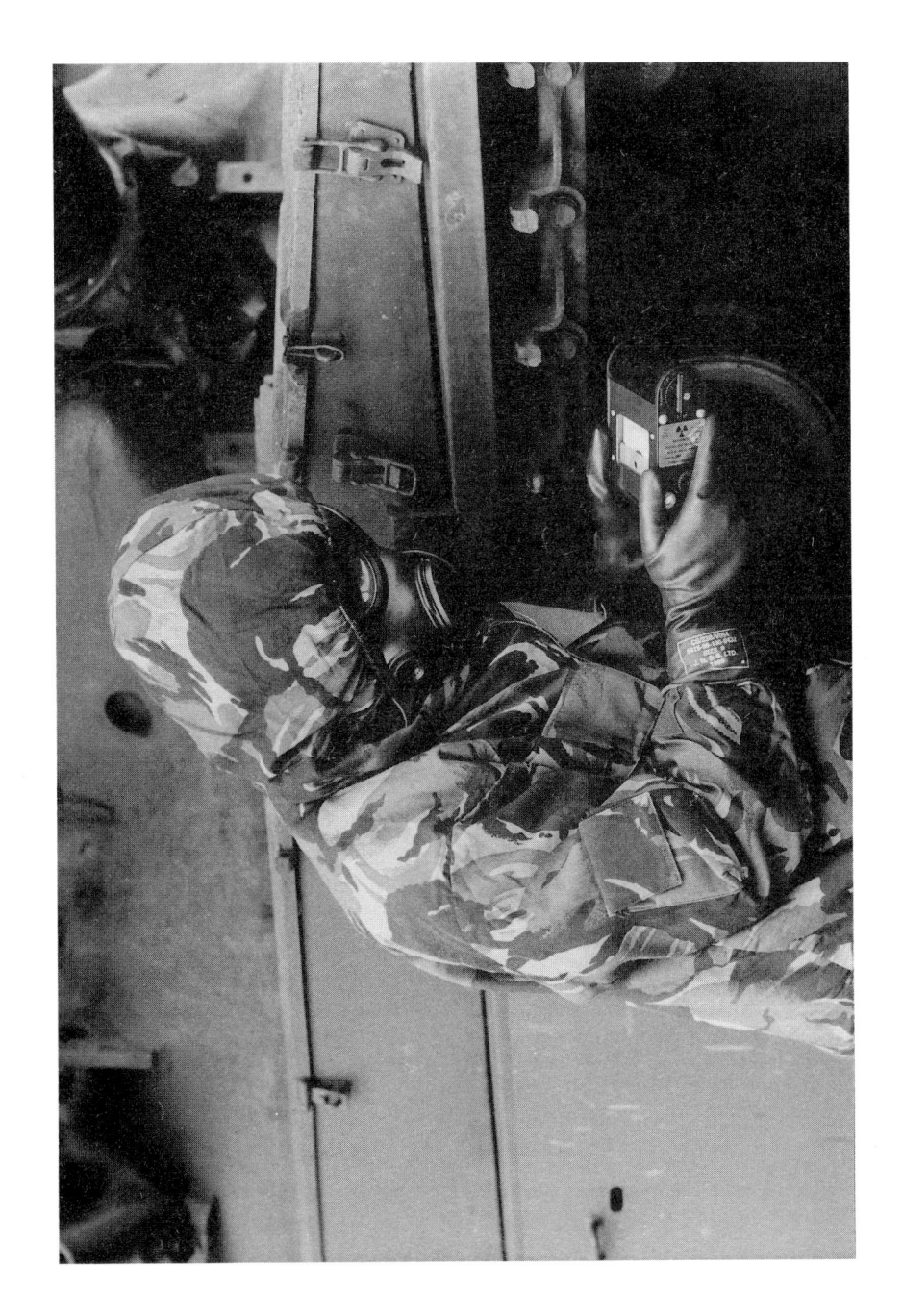

Fig. 6.5 The author using a portable dose rate meter for contamination measurement (*Author*)

are also produced for fitting in helicopters and drones. Since they average the dose rate over quite a large area they may fail to detect local 'hotspots' caused by rain or the local topography.

Contamination Meters

Troops who have to work in an area where there is fallout obviously need a dose rate meter to warn them of the hazard from fallout on the ground. They also need another, rather more sensitive instrument which can be used to detect areas of skin, clothing, vehicles and equipment which are contaminated by fallout. Ideally such a contamination meter responds only to beta particles, because their short range in air compared to that of gamma rays means that the reading is then determined by the contamination on the surface being checked – general contamination on the ground will not have much effect.

Various forms exist, but the simplest consists of an ionization chamber with a thin foil window which is normally covered. It then behaves as an ordinary dose rate meter. However, if the window (which is thin enough to allow beta particles through) is uncovered, the reading will be dominated by beta particles coming from surfaces close to the window. Figure 6.5 shows an instrument of this type. Contamination meters are necessary, not only to warn of the presence of contamination, but also to check that decontamination procedures have in fact removed the fallout.

Protection against Radiation

There are a number of steps which can be taken to reduce the effects of radiation on man and equipment. The most important are the reduction of exposure time, increasing one's distance from the source of radiation, and the use of shielding. All these aim to reduce the dose which men or equipment receive. There is, for man, a fourth approach – chemical prophylaxis – the administration of drugs which reduce the effects of radiation.

Exposure Time

Immediate nuclear radiation is, as we have seen, delivered quite quickly. However, the drills recommended for men caught in the open to reduce blast and thermal effects, such as falling into a ditch or slit trench, may be of some benefit, particularly for the higher weapon yields. The benefits of rapidly moving out of an area of high residual radiation are obvious.

Distance

As far as immediate radiation is concerned the only practical ways of increasing the distance from a nuclear strike are to disperse troops beforehand as widely as the military situation permits, and the obvious one of keeping clear

FIG. 6.6 Rapid decontamination of vehicles (*Lft Col A.P. Farquar*)

of potential targets. Much more can be done when the problem is residual radiation. Moving out of the affected area is sensible when this is possible. Failing that, we can reduce the dose to men by removing the radioactivity. Scraping contaminated topsoil a few metres away from a slit trench would reduce the dose to its occupants by about 30 per cent.

Decontamination

This is a very effective method for increasing the distance between man and radioactivity. Fallout can be removed from most equipment by hosing down with water, scrubbing and using a detergent if one is available. Figures 6.6 and 6.7 show two approaches to vehicle decontamination. For delicate equipment, vacuum cleaning is the best approach. The human body can be decontaminated by washing with plenty of water. Tinned or packaged foods can be decontaminated by simply cleaning the outer surface of the pack; for exposed foods one must cut a thin slice off all exposed surfaces. Filtration of water for drinking, using standard army equipment, considerably reduces the radioactivity level, although if the gamma dose rate locally is low enough to allow continuous occupation, the water in the area can usually be drunk in normal quantities.

Filtration of air is essential to prevent the inhalation of fallout particles. Troops working outdoors should wear full NBC clothing and respirators, as shown in the frontispiece. Vehicles and other forms of collective protection should have all doors and hatches closed, and their air filtration systems should be operated.

One must bear in mind that if the residual activity of equipment is due to

FIG. 6.7 Decontamination by scrubbing (*Lft Col A.P. Farquar*)

neutron-induced activity, decontamination will be useless since in this case the activity is not simply lying loose on the surface but comes from the actual atoms within the affected equipment. The only thing one can do then is to wait for the activity to decay of its own accord.

Shielding against Gamma Rays

The process by which gamma rays give up their energy as they pass through matter was discussed earlier in this chapter. There we saw that it is the weight which determines the effectiveness of a material as a gamma-ray shield. Listed below are the half-thicknesses for immediate gamma rays of common shielding materials, with the weight of a square metre of each:

steel: 2.8 cm,   228 kg/m^2
concrete: 10 cm,   241 kg/m^2
rammed earth: 15 cm,   245 kg/m^2

Weights and thicknesses for the less energetic, so less penetrating, residual gamma rays would be about two-thirds of those given above.

These figures show that there is not much we can do to provide shielding for men and equipment other than that already available in the form of trenches, dugouts and vehicle armour. For example, to reduce by half the gamma dose to a small electronic item in the form of a 30 cm cube, a steel shield would need to be 2.8 cm thick and would weigh over 100 kg. To reduce the dose to one-tenth, its weight and thickness would be more than three times these values.

The effectiveness of a shield is specified by its transmission factor, which is the ratio of the dose (or dose rate) inside a shield to the dose (or dose rate) outside. Sometimes its reciprocal, called the protection factor, is used instead. Some values are quoted in Table 6.4, but these should be taken as only a guide. It is important to appreciate that for armoured vehicles it is the thickness of the thinner regions, such as the roof and the floor, which most influence the transmission factor for the vehicle as a whole.

TABLE 6.4
TRANSMISSION FACTORS FOR NUCLEAR RADIATION

Shield	Radiation		
	Immediate Gamma	Neutron	Residual Gamma
open trench	0.2	0.3	0.1
shelter with 0.9 m earth cover	0.007–0.02	0.01–0.05	0.001–0.005
building basement	0.1–0.6	0.1–0.8	0.05–0.1
tank	0.2	0.3	0.06
infantry armoured vehicle	0.5–1.0	0.6–1.0	0.2–0.5
soft vehicle or aircraft	1.0	1.0	0.5
ship	0.001–1.0	0.001–1.0	0.001–0.5

Shielding against Neutrons

Shielding against neutrons is a more complex task than shielding against gamma rays, and is usually less effective as the figures in Table 6.4 show. The interactions neutrons undergo with materials were discussed earlier in this chapter; many result in the emission of gamma rays, which must in turn be absorbed. Hydrogenous materials, such as damp earth, concrete and plastics, are quite good neutron shields since not many gammas are produced by the neutrons as they are slowed. However, when neutrons are captured rather energetic gamma rays are emitted, partially cancelling the benefit of removing the neutrons. The addition of quite small amounts of boron is beneficial since it captures neutrons very readily, releasing only a few low-energy gamma rays. If the incorporation of boron is unacceptable, a thin layer of a boron compound on both sides of a hydrogenous shield is of some benefit.

Metals such as steel are far from ideal because of the gamma rays produced when neutrons are slowed and captured. Useful improvements may be made by adding a hydrogenous layer, preferably containing boron, to both sides of it. Reactive armour on the exterior of a tank and a spall liner on the interior both reduce the neutron-transmission factor.

Chemical Prophylaxis

The early somatic effects on man which concern us are nausea and vomiting, tissue damage, and the general incapacitation induced by higher doses. All these effects can be reduced to a greater or lesser extent by the administration of appropriate drugs. This is called chemical prophylaxis or therapy. However, there are problems. For example, drugs which reduce the incidence of incapacitation if they are taken 30 minutes before irradiation have the side effect of inducing vomiting. The same drugs also reduce tissue damage, but the maximum tolerable dose gives only a low dose reduction factor of about 1.2 to 1.5. Further, many drugs which reduce nausea and vomiting are known, but all have the undesirable side effect of acting as sedatives.

Conclusion

Nuclear radiation has presented us with rather more difficulties than the thermal and blast we have considered in earlier chapters. First, it is a novel, invisible, and potentially lethal threat to man, and can also damage his electronic equipment. Secondly, in addition to the immediate radiation, we have to be prepared to face residual radiation which may be at dangerous levels for days or even weeks after a burst. It is essential that troops understand these dangers and what they can do to mitigate them by shielding and decontamination. If they do, then their chances of remaining effective after a strike will be greatly enhanced.

7.
Electromagnetic Effects

Introduction

The nuclear weapon produces three primary outputs which are fairly widely known: thermal, blast and nuclear radiation. Each of them can produce casualties among troops and damage to equipment out to ranges of a few kilometres; these are the effects which tend to spring to mind when we think of nuclear weapons. However, they are not the whole story, for there is another set of effects – the electromagnetic effects – which, while they pose no direct threat to man, may be even more devastating to equipment.

There are actually two separate effects we have to consider. The first, and by far the more important, is the electromagnetic pulse (EMP) – an intense pulse of radiofrequency energy which is generated when a nuclear weapon is detonated on or above the earth's surface. It is not a primary output of the weapon itself, but is a consequence of gamma-ray interactions with the atmosphere.

EMP can cause damage or upsets to unprotected electronic equipment out to ranges of thousands of kilometres, so one suitably placed weapon could disrupt many human activities over a whole continent. Most of this chapter is devoted to the generation of EMP, its nature, how it affects systems, and the steps we must take in equipment design and operation to lessen the risks of damage.

The second electromagnetic effect is not on equipment, but on the propagation medium for radio waves, which is the atmosphere. Atmospheric ionization may affect transmission in some frequency bands for many hours after a nuclear burst. This chapter has a summary of these effects on propagation.

EMP Origin and Nature

The generation process and the electromagnetic fields which result are much influenced by the burst height. We must look separately at two cases – the endo-atmospheric burst (within the atmosphere) with a burst height less than 30 km and the exo-atmospheric burst, at above 40 km.

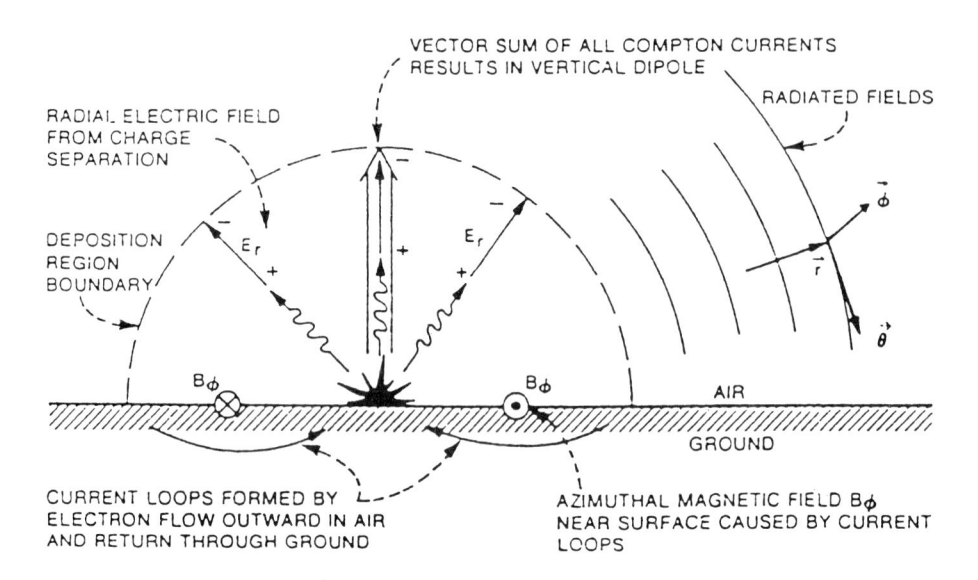

FIG. 7.1 EMP from a surface burst

Endo-atmospheric Burst

We shall look first at a burst a few kilometres above the ground. The immediate gamma rays we discussed in Chapter 6 stream out in all directions and interact by Compton scattering with atoms in the atmosphere. In this process they eject electrons which travel outwards a few metres at most from their parent atoms (which are now a positive ions). This outward radial movement of electrons constitutes an electric current, called a Compton current. It creates a separation of electric charges (the positive ions and the electrons), so a radial electric field is set up, directed outward from the burst point. Few of the gamma photons will penetrate more than 2 or 3 km, so most of the charge separation occurs within a sphere of this radius round the detonation. This is called the source region or deposition region. It is shown in Figure 7.1 (which actually shows a surface burst rather than the mid-altitude one we are discussing).

The enormous intensity of the immediate gamma pulse means that the electric field reaches a very high value – several tens of kilovolts per metre (kV/m) in a few nanoseconds. Figure 7.2 shows how it varies with time. The peak field occurs when the charge separation is at its maximum. Thereafter the electric field urges the electrons back towards their parent ions, with which they eventually recombine. This rather slower process is responsible for the long plateau and tail of the electric field pulse shown in Figure 7.2.

The laws of electromagnetism known as Maxwell's equations tell us that wherever there are currents or electric fields which change with time, there will also be magnetic fields. The Compton current in the source region

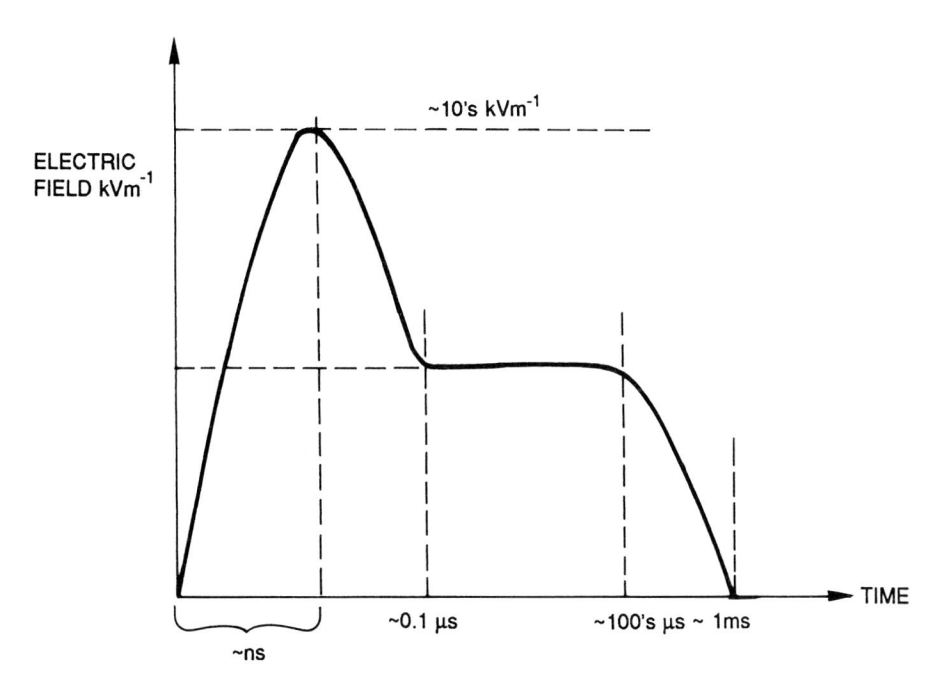

FIG. 7.2 Endo-atmospheric pulse

therefore generates a magnetic field, which rises to a peak value of about 100 amperes per metre just as rapidly as the electric field, but the time for which it persists is shorter.

These intense, transient fields in the source region (where actual currents are flowing) do not constitute a radiated electromagnetic wave. However, the Compton currents will radiate waves just as currents flowing in metallic conductors do. For the mid-air burst we are considering the radiation from the Compton currents flowing upwards from the burst point is almost exactly cancelled by radiation from currents flowing downward, so there is only a small radiated wave outside the source region. It weakens as it spreads, and would be unlikely to cause damage on the ground.

Surface Burst

If our endo-atmospheric burst is not at mid-altitude but takes place on or near the surface (the more likely scenario), then we have the situation shown in Figure 7.1. As before, there will be strong outward Compton currents of electrons, but many of them will flow back towards the burst point through the conducting ground, so forming current loops which enhance the magnetic field in the source region. However, in this case neither the downward-travelling gamma rays nor their Compton electrons can travel far into the ground, which in any case is a conductor, so outside the source region the

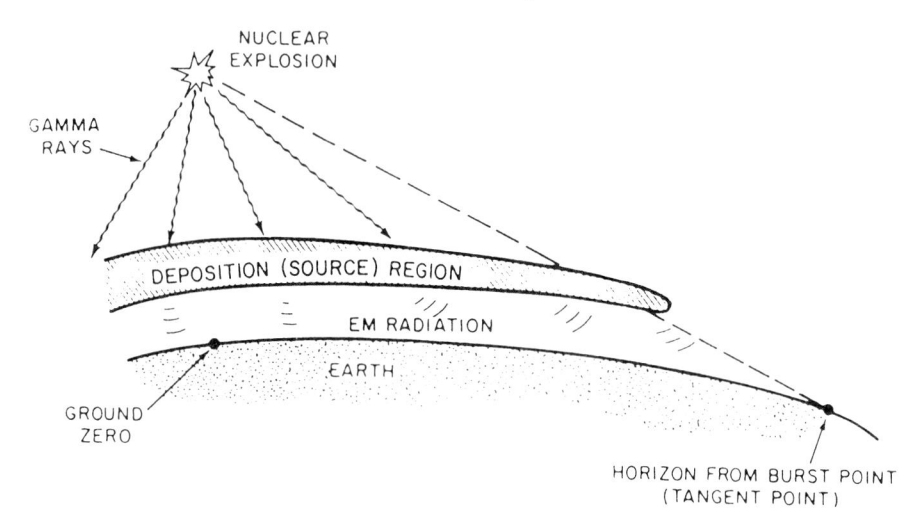

```
                        NUCLEAR
                        EXPLOSION
GAMMA
RAYS

         DEPOSITION (SOURCE) REGION
                   EM RADIATION
              EARTH
 GROUND
 ZERO
                              HORIZON FROM BURST POINT
                                   (TANGENT POINT)
```

FIG. 7.3 Generation of exo-atmospheric EMP
(*US Department of Defense – Department of Energy*)

electromagnetic radiation from the upward and the downward Compton current no longer cancel out. There will therefore be radiated fields, but the field strengths are well below those within the source region, which are the principal threat to electrical systems.

Exo-atmospheric Burst

When the burst height is 40 km or more it is still electrons ejected by Compton scattering which generate the pulse, but the geometry, which is shown in Figure 7.3, is different. Gamma rays travelling downward encounter few air molecules until they reach the upper tenuous layers of the atmosphere at about 40 km above the ground. Most will undergo Compton scattering in a source region between 20 and 40 km up, so we again have a Compton current of electrons. However, in the rarified air at these altitudes each electron can travel hundreds of metres. Their trajectories are now influenced by the earth's magnetic field; they move in helical paths with a radius of curvature of a few hundred metres. This is shown in Figure 7.4, but with artistic licence. Each electron, in fact, makes less than a single turn of a helix before it comes to rest.

The consequence of these mechanisms is that there is a transverse component to the Compton current, which can be shown to result in the radiation of electromagnetic waves in the original direction of travel of each electron. Since gamma rays and radio waves travel at the same velocity, all the sources on a line between the burst point and a point on the ground produce radiation which reaches the ground in phase. Therefore all the sources reinforce one another, producing very high field strengths in the radiated EMP reaching the ground.

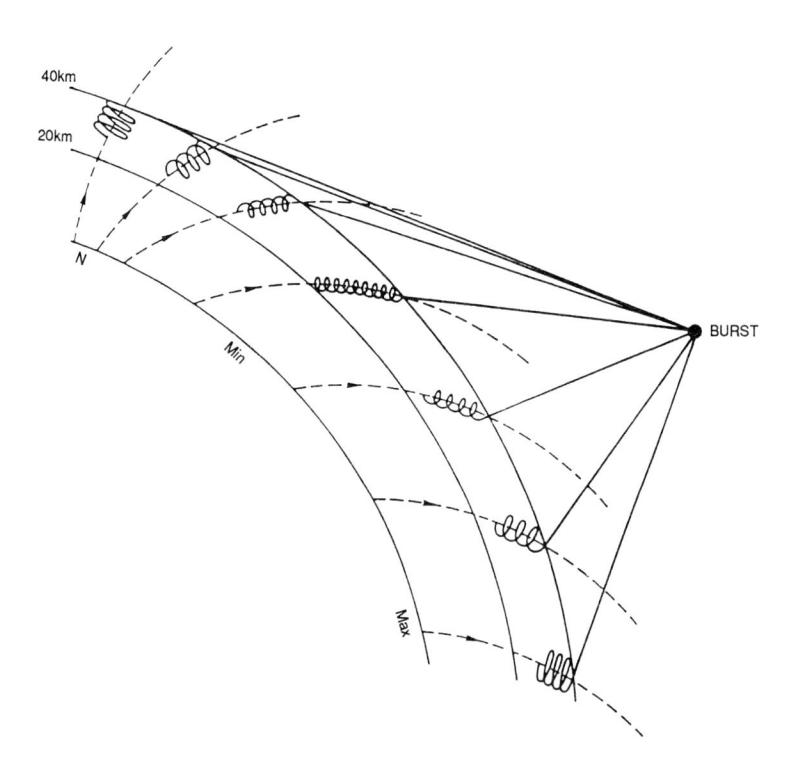

FIG. 7.4 Role of magnetic field in Exo-atmospheric EMP generation

The shape of the pulse seen on the ground is shown in Figure 7.5. The rise time of a few nanoseconds and the peak electric field of about 50 kV/m are similar to those in the endo-EMP, but since we are dealing with a radiated wave there is now a fixed relation between the electric field E and the magnetic field H:

$E/H = Z_0 = 120\pi$ ohms

where Z_0 is the electromagnetic impedance of free space. The peak magnetic field is therefore about 130 amps/metre.

In the low air density of the source region the Compton electrons have an easier return passage to their parent atoms than do those in the endo-atmospheric case, so the field only persists at a high value for less than one microsecond, as Figure 7.5 indicates, but the 'early time' pulse it shows is not the end of the story. Gamma rays emitted later in the immediate radiation, together with those which have been scattered *en route*, continue to contribute to the EMP for a few milliseconds, but their contributions are not in step,

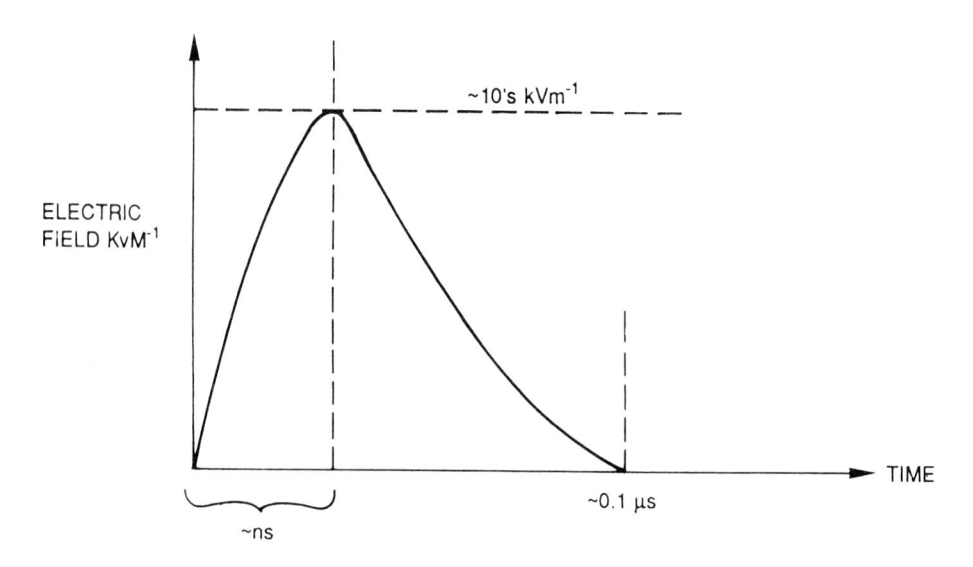

FIG. 7.5 Exo-atmospheric EMP: early time

so fields in this 'mid time' portion of the pulse are only a few hundred volts per metre – too low to show in Figure 7.5.

The exo-atmospheric EMP story is not over yet. The weapon debris forms a highly conducting plasma. As it expands it pushes aside the earth's magnetic field lines, creating ripples in them called magnetohydrodynamic waves. These propagate around the earth, and, though their amplitude is only a fraction of a volt per metre, they may persist for up to 100 seconds. This is the 'late time' component; like the mid-time component it is too low in amplitude and frequency to have much effect on military systems, but it can generate large, almost unidirectional currents in very extensive systems, such as national power distribution networks, in much the same way that magnetic storms do in areas near the earth's magnetic poles, with serious consequences on components such as power transformers.

Characteristics of Exo-atmospheric EMP

The most striking feature is the enormous area exposed to damaging field levels, as we might infer from Figures 7.3 and 7.4. Equipment is at risk out to the visible horizon as viewed from the burst point. When the burst height is 400 km the affected area extends out to 2,200 km in all directions from GZ (Figure 7.6), so one high-altitude burst can damage electrical and electronic systems over a whole continent. The peak fields from a surface or low air burst may be somewhat higher, and may persist at a high level for longer (implying that there is more energy associated with it), but these high levels extend only out to a few kilometres from GZ, affecting an area where blast, thermal and nuclear radiation will, in any case, have caused damage. By

Electromagnetic Effects

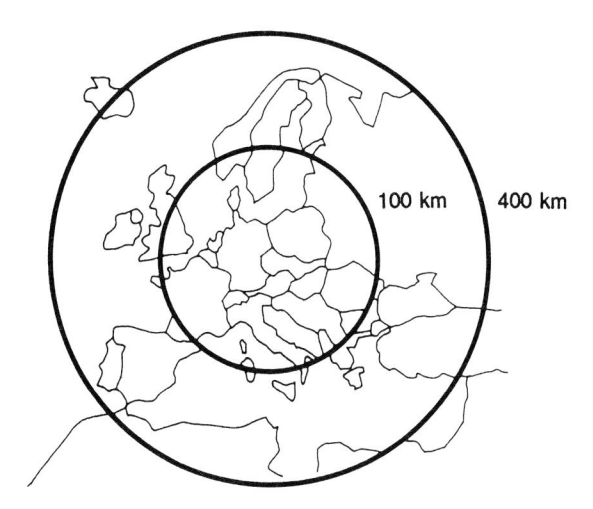

FIG. 7.6 Area affected by exo-atmospheric EMP; heights of burst 100 km and 400 km

contrast, the EMP from a single high-altitude burst can disrupt an entire national power or communications system. Indeed, EMP is the only significant effect on the ground for a high-altitude burst, even though the EMP energy amounts to little more than 0.01 per cent of the total weapon output.

When we discussed the other weapon outputs, data about the output for a 1-kT weapon were given, together with scaling laws which enable one to deduce the output for any other yield. In our discussion of EMP we have made no mention of weapon yield so far. The reason is that when intense nuclear radiation pulses produce ionization in air there are saturation effects which make the resulting EMP fields not fully dependent on yield. Details of field strengths and pulse shapes quoted earlier may therefore be taken as applying for all yields.

The exo-atmospheric pulse, like any transient, has a frequency spectrum which in this case extends up to hundreds of megahertz (Figure 7.7(a)). It is seen to contain significant energy in almost all the frequency bands used for radio communication and surveillance. The total EMP falling on each square metre of the area it affects is about one joule, which is not a great amount (it is about the same as the kinetic energy a small apple has after falling one metre). However, as we shall see, it is enough to damage up to ten million semiconductor devices.

We have been looking at the EMP experienced on the earth's surface. An aircraft will suffer similar fields, but a satellite, being above the source region, will receive only a much weaker pulse reflected from the earth's surface. System-generated EMP, discussed later, is likely to be a greater problem for the satellite.

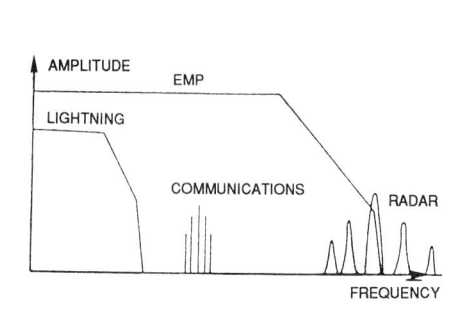

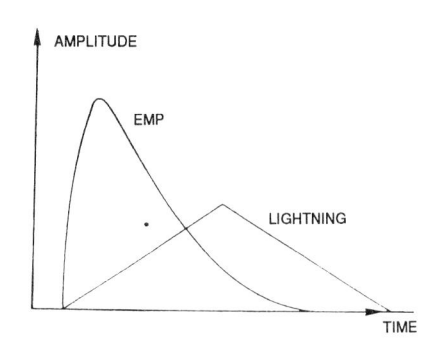

FIG. 7.7 (a) Spectrum comparison (b) Comparison with lightning

Comparison with Other Electromagnetic Fields

Military electronic equipment is designed to be operable close to radio and radar transmitters, which could expose it to fields of tens or hundreds of volts per metre, without suffering damage. This is unlikely to guarantee survival when it is exposed to EMP fields of 50,000 V/m.

Many electrical and electronic systems incorporate protection against lightning strikes. Since a direct strike causes a current much larger than that which EMP would induce, one might suppose that the incorporation of lightning protection would be more than is needed for protection against EMP. This is not so: the EMP pulse rises very rapidly, in a few nanoseconds, but a lightning discharge is usually much slower (Figure 7.7(b)). Some of the protective devices we shall be discussing, while fast enough to deal with lightning strikes, will not act sufficiently quickly to protect against EMP.

System-generated EMP

So far we have been concerned with EMP generated by weapon gamma rays interacting with the atmosphere. Very similar effects may occur if an equipment is itself exposed to an intense pulse of nuclear radiation. It interacts with the structural materials of the equipment by ejecting electrons, so giving rise to charge separation and the creation of intense and potentially damaging electric and magnetic fields within the equipment. These are called system-generated EMP (SGEMP).

The situation most likely to produce serious SGEMP is an exo-atmospheric burst and a target such as a satellite or missile which is also above the atmosphere. The radiation then consists of the immediate gamma rays with, in addition, the enormous number of low energy X-rays which are a feature of such bursts (Chapter 3). Since there is little absorption, radiation intensities are high enough to produce serious SGEMP problems even at ranges of several tens of kilometres. SGEMP could also be significant from low air bursts of low yield, but in this case only gamma rays would contribute to it.

Electromagnetic Effects

The prediction of SGEMP fields and effects is complex, and has to be done separately for every type of system, since the results are strongly dependent on the materials involved and their distribution round the system. The usual approach is therefore a practical one: a simulator is used to irradiate the system, and the fields induced and the damage they do are measured.

EMP Damage and Coupling

The high fields associated with all three types of EMP induce high voltages and large currents in any electrical conductors they encounter, often with harmful results to devices connected to the conductors. Here it is important to note that EMP energy collection is not confined to conductors which are deliberately designed to abstract energy from electromagnetic waves, such as antennae. Therefore EMP-induced currents will also flow in power lines, telephone lines, data cables, and also in metal water pipes or water flowing in a plastic pipe. Fibre-optic cables, of course, are immune to EMP effects (unless they have a metal strength member), and so provide an easy solution to some EMP problems.

Damage

EMP is unique among the weapon outputs in that it alone poses no significant threat to man. Most at risk are the semiconductor devices which are at the heart of all our electronics. The problem arises not from the very small energy which EMP deposits in a device direct, but from the currents it induces in conductors connected to it. With most devices the problem is due to the currents flowing through the junctions within the devices. This causes local overheating and permanent parameter changes, but other effects such as the burnout of metal interconnections are possible. Typical energies to cause damage are listed in Table 7.1. It shows that some light electrical devices, such as detonators, relays and motors, are also susceptible, though energies sufficient to cause damage are larger. There is also the possibility of EMP-induced voltages causing insulation breakdown in cables and capacitors.

Even if an EMP-induced transient is much too small to damage components in a system, it can still upset its operation. Transient malfunction is most likely in digital logic circuits, so that computers and their memories and microprocessors will all be vulnerable. Transients with energies as low as 10^{-9} joule may be all that is needed to cause these upsets.

Coupling

It is not a simple matter to calculate the current which EMP will induce in a long conductor. The precise value depends on the polarization of the EMP (for the most important exo-atmospheric EMP it is horizontal), the cable's orientation relative to the EMP, its length, whether it is on the ground, raised, or

Nuclear Weapons

TABLE 7.1
EMP DAMAGE ENERGIES

Device	Energy (joules)
microwave diode	10^{-7}
FET	2×10^{-5}
input gate IC	8×10^{-5}
audio transistor	4×10^{-3} to 7×10^{-2}
explosive bolt	3×10^{-4}
relay	2×10^{-3} to 10^{-1}
vacuum valve	1
electric motor	10^{3}

buried, and also the electrical conductivity of the ground under the cable. This last determines how well the ground reflects the EMP, the induced current being greatest when the ground is a poor conductor.

One must expect the current which an exo-atmospheric burst induces in a long cable to rise to a few thousand amperes in a few hundred nanoseconds. Fortunately the rate of rise of the current is much slower than that of the EMP fields, but it is still much faster than current caused by lightning.

For short conductors a few metres at most in length simplications may be made; the peak current is given by:

$$I_{pk} = 20 \, l^2 \text{ amp}$$

where l is its length in metres. Just as the electric field couples to straight conductors, so the magnetic field couples into loops, and it is easy to deduce that a loop of one square metre in area could have about 30 kV induced in it.

Many systems, even quite small ones such as that in Figure 7.8, will have several paths for EMP entry. Obviously it will couple into any antennae (*a*), but also into any incoming cables such as the data cable (*b*) and earth connections (*c*). The skin may not behave as a perfect shield – indeed, if there are open hatches (*d*) or poor joints (*e*) there will be appreciable fields inside which will couple into internal wiring. It is important to realise that once EMP-induced currents are in the system they will diffuse through it by coupling between cables (*f*), so all parts of the system will be at risk.

Here it may be helpful to point out that very small systems, such as a modern wristwatch or a pocket calculator, do not have any conductors long enough or loops large enough to pick up damaging amounts of energy. However, a man-pack radio with a whip antenna and handset leads would be potentially vulnerable unless it had been hardened against EMP.

Protection against EMP

The simplest as well as the most obvious way of protecting a system against EMP is to put it inside an electromagnetic shield, called a 'global shield'.

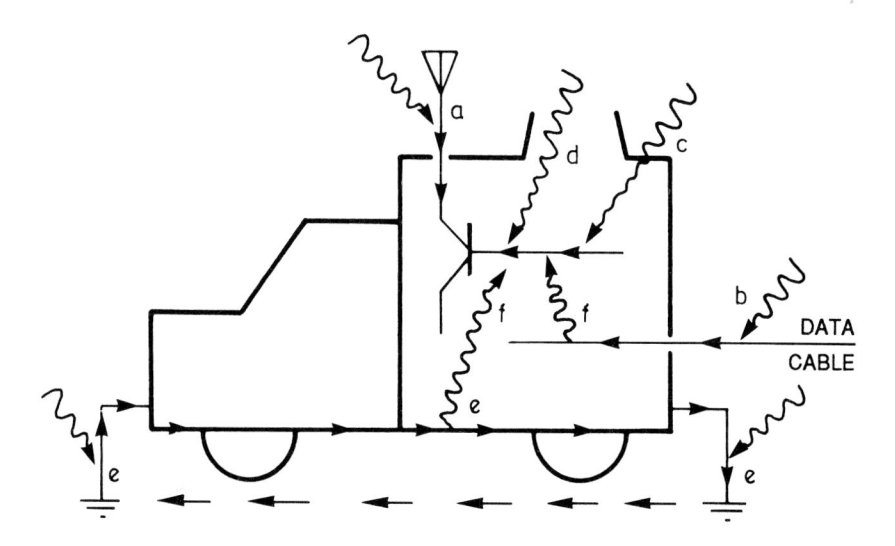

FIG. 7.8 Entry paths for EMP

This may be made of any reasonably good conductor such as steel, aluminium or copper, and need not be very thick – 3 mm of steel is more than adequate. However, it is essential that it should be free of gaps and apertures through which EMP energy could enter.

This approach may be used for any size of system – a manportable instrument, the hull of an armoured vehicle, tank or ship, or a large underground headquarters. If one ensures that the shield is able to reduce the EMP fields inside to about 1 volt/metre, then within the shield no special precautions need to be taken. Indeed, ordinary commercial equipment, not hardened in any way against EMP, may be installed inside it. The fuselage of an aircraft is of little use as a shield because of large apertures such as the cockpit, so electronic units for use in aircraft need to be individually protected.

Apertures

All systems need apertures. A man-portable instrument must have a removable cover for the battery compartment, and probably a panel which can be removed for servicing and testing. A larger system, such as a tank, ship or underground headquarters also needs doors or hatches for its crew. All these must be treated so that when they are closed they are electrically continuous with the global shield. This can be achieved by fitting them with conducting gaskets which may be wire mesh, conducting rubber or metal finger stock.

Apertures such as air inlets and outlets (which must always be open) are also needed. The usual way of treating them is to cover them with a metal honeycomb grille with the individual cells only a centimetre or so across but of a depth several times this. Each cell acts as a waveguide, but because of its

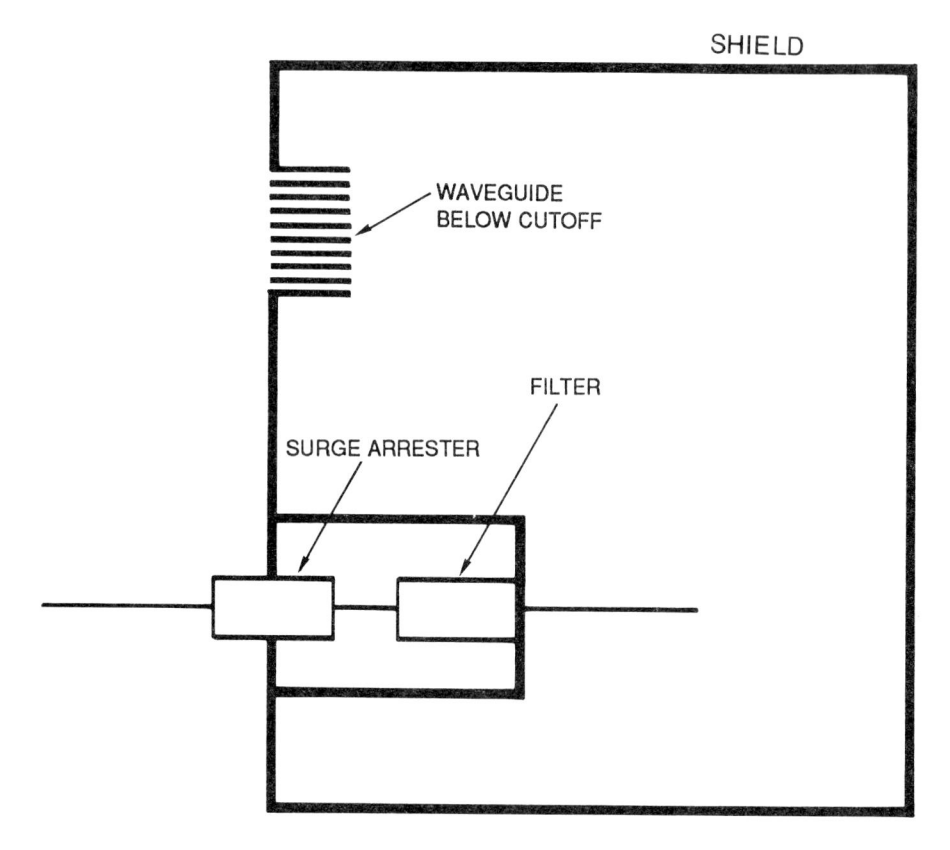

FIG. 7.9 Global shield with waveguide below cutoff

small size it has a cutoff frequency (the lowest frequency which can pass through it) in the GHz region. The amount of EMP energy at or above this frequency, and so able to penetrate the grille, is negligible. Figure 7.9 shows a global shield with a grille acting as a 'waveguide below cutoff'.

Cables

Nearly all systems have cables which must penetrate the global shield. These may be telephone lines, mains power leads, antenna connections, data cables, and so on. EMP will induce currents in them; it is clear that these currents must not be allowed to penetrate the shield, for if they did they would spread to all the equipment and circuits within it, leading to multiple failures. If the cables are not too long and link systems which have their own global shields, the solution is to use shielded cable. The cable shield then extends and links the two global shields. Solid metal conduit performs better than the braided screens usually found in flexible cables, but whatever material is used, it must be well connected to both global shields.

Electromagnetic Effects

For unshielded cables, one must fit a protection system to each conductor at its point of entry to the global shield, using the devices described in the next section. However, fibre-optic cables may simply be fed in through a small metal tube acting as a waveguide below cutoff.

Protective Devices

The commonest form of protection for cables consists of a surge arrestor or clamp between each conductor and earth. These are devices which normally behave like an open circuit. However, if a transient causes the voltage across them to exceed a certain value, they change to a conducting state and then behave like a short circuit, reflecting the transient back down the cable.

The most popular clamps are spark gaps, transient absorbing zener diodes, and metal oxide varistors. None are ideal in all respects – for example, spark gaps are not necessarily very fast-acting, and the voltage across them may reach quite a high value before they begin to conduct. A hybrid arrangement of two devices is therefore often needed; a spark gap followed by a low pass filter is common (Figures 7.9 and 7.10); the filter absorbs the transient which gets past the spark gap before it begins to conduct.

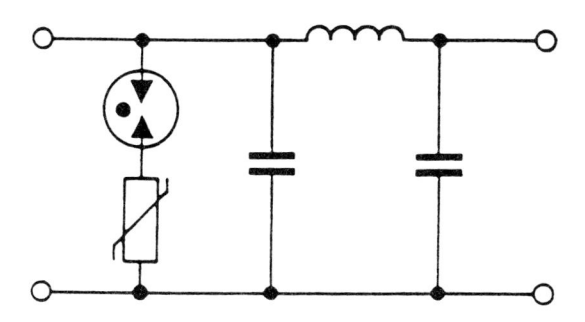

FIG. 7.10 Hybrid protection

Precautions in Use

The ability of an equipment to survive the blast, thermal and nuclear radiation effects of a weapon is largely determined by the original design and the materials and components used in it. Survivability will not necessarily be affected by poor or inadequate maintenance. Things are different for EMP. Damage to gaskets around doors and hatches, or an open hatch, may jeopardise the EMP survivability of a system, as may drilling a hole in the shield and feeding an unprotected cable through it. It is therefore important that all personnel concerned with a system should not only have a good understanding of the EMP threat to it, but also should be trained to watch out for poor maintenance and modifications which could affect EMP survivability, and to take prompt action when any are discovered.

Many modern practices are not necessarily compatible with EMP survivability. The radio village concept, adopted by many armies for their higher HQs, with its long radio and generator leads and remote antenna connections, not to mention remote handset arrangements, will be at great risk unless comprehensive hardening against EMP has been carried out. Good maintenance of cables and connectors is essential, and careful cable and earth leads layout, ensuring that minimum lengths are used and loops avoided, will reduce EMP pickup.

When a warning of our own nuclear strikes is given, or when the probability of enemy strikes is high, there are simple measures which will greatly reduce the dangers. Upon receiving the warning, antennae should be taken down or disconnected, cables unplugged from vehicles and equipment, and doors, hatches and windows in vehicles and buildings checked for proper closure. These steps will go a long way towards preserving equipment.

Effects on Radio Propagation

In addition to the EMP effects we have been discussing, weapon bursts produce a large amount of ionization in the atmosphere and changes in the ionized layers above it. These may affect sky wave propagation for some time after a burst. The scale of the effects depends markedly on the weapon yield and burst height; generally the higher the burst the more pronounced and long-lasting are the effects. The ground wave is unaffected, as is direct line-of-sight propagation (unless the line happens to pass through a region of enhanced ionization, such as the cloud of radioactive debris). The sky wave suffers most.

The effects extend to a distance of from hundreds to thousands of kilometres from GZ, and last for a time of from minutes to hours, except where noted below. As they are strongly frequency-dependent we must look at each band separately:

VLF:	lowering of reflection height causes phase and amplitude changes to sky wave modes;
LF and MF:	absorption and defocusing of sky waves;
HF:	absorption of sky waves, particularly by day; new propagation modes may appear; multipath interference;
VHF:	absorption and multipath interference up to hundreds of miles, persisting for up to tens of minutes;
UHF:	absorption for line-of-sight propagation through highly-ionized regions for a few minutes.

Conclusion

EMP probably presents the greatest threat to both civil and military systems (virtually all of which depend on electronics), since a single exo-atmospheric burst can have continent-wide effects. Equipment can be hardened against EMP, but the level of protection tends to degrade with time. Therefore high standards of personnel training and equipment maintenance are essential to maximize the chances of our systems surviving EMP.

8.
Nuclear Survivability

Introduction

Before the introduction of nuclear weapons it was generally true that, with the battlefield weapons then employed, much equipment had a better chance of surviving than had the soldiers operating it. This was still the case in the early days of nuclear weapons, when it was sometimes said, perhaps rather optimistically, that if equipment were soldier-proof it would also probably be atomic-bomb-proof. Staff Requirements in those days merely stated that equipment 'should be resistant to the effects of nuclear weapons'. This placed few constraints on the designer, since any equipment will survive a nuclear burst if the yield is small enough, or if the burst point is far enough away.

A radical rethink was needed when solid-state electronic devices came into widespread use in the 1960s, since it was soon discovered that many of them, unlike the electronic vacuum tubes they replaced, were likely to be upset or damaged by nuclear radiation doses which a man might survive. At around the same time, the threat which EMP presented to the new devices also became apparent. It was the realization of the potential seriousness of these effects which led to the introduction by the Army in the following decades of a more quantitative approach to nuclear survivability, which may be defined as 'the capability of a system to withstand a nuclear environment without loss of its ability to accomplish its designated mission'.

In this chapter we explain the basis for deciding whether nuclear survivability should be specified for a particular equipment, and the procedure for determining the levels of nuclear-weapon effects (called 'survivability criteria') which it should be able to withstand. We also include some general guidance for project managers and designers to help them produce nuclear-survivable equipment.

UK Survivability Policy

It is the policy of the British Ministry of Defence that Staff Targets and Requirements should state whether nuclear survivability is required for a particular system or equipment. If it is not needed, the justification for the decision to exclude it must be given. These decisions are made by the appropriate Operational Requirements Branch in consultation with the users. If nuclear survivability is needed the nuclear environment is to be clearly specified as a set of survivability criteria detailing the precise levels of thermal radiation, blast, nuclear radiation and EMP which the equipment must withstand.

The tendency towards the proliferation of nuclear weapons means that there is a continuing risk of their being employed strategically and also as tactical battlefield weapons. The current hardening philosophy is that if 50 per cent or more of a unit's men are fit to fight after a nuclear strike, then any of their equipment which is vital for post-strike operations must have a high probability of remaining fit for use. The reasoning behind this is that, on average, a unit suffering 50 per cent casualties will still just be able to complete its immediate mission. Most current UK and NATO survivability criteria are based on this philosophy, but in exceptional circumstances criteria may be based on other casualty percentages. The aim of the designer must be to balance the survivability of the equipment against the vulnerability of its operators. It would clearly be uneconomic to overharden it against one weapon effect while leaving it vulnerable to another: balanced survivability must be the aim.

The Decision to Specify Nuclear Survivability

It is obvious that potential aggressors, and our own soldiers as well, should all know that our equipment is hardened against nuclear-weapon effects, since this knowledge not only improves the morale of our troops, but also increases the value of our forces and equipment as a deterrent to aggression. There are also moral as well as practical arguments which justify the nuclear hardening of our equipment. Soldiers who survive a nuclear attack have a right to expect that if they are fit to fight their equipment should still be operable.

Current UK policy is to produce equipment of high quality but in limited quantities. Inevitably the unit cost is high and only a small number of replacement equipments is provided. In any case, under the conditions of nuclear war, replacement in the field may be difficult or impossible. For these reasons it is essential that most of our equipment should have a high probability of surviving when there are sufficient crew left to operate it.

Nuclear hardening inevitably involves some expenditure of time and resources, though the additional effort need not be great if hardening is considered from the earliest stages and if those involved have an understanding of nuclear-weapon effects. Nevertheless, the cost of hardening means that the following questions should be asked before the decision to specify nuclear survivability is taken:

> Is the equipment likely to be used in a nuclear theatre, and, if so, is it vital for post-strike operations?
>
> Can it be quickly and easily replaced under nuclear war conditions if it does not survive?

Is the equipment going to be used only in conjunction with old equipment, the survivability of which is unlikely or unknown? If so, is the old equipment likely to be replaced by an equipment for which nuclear survivability might be specified?

When equipment is being designed to meet a purely military requirement, the specifying of nuclear survivability will usually be found to be necessary and worthwhile. In the case of equipment produced primarily for a civilian market and procured 'off the shelf' it may not be economical to specify full nuclear survivability, though it may be possible to improve the hardness overall by the simple modification of particularly vulnerable components, especially if they are going to be modified for other reasons for military use. If the decision is made not to specify nuclear survivability the consequences must be understood; it may mean an increase in the provisioning of spare components or in the scaling of complete replacement equipments.

Definition of the Nuclear Environment

The emphasis in survivability is strongly on the immediate outputs from a weapon (thermal radiation, blast, immediate nuclear radiation and EMP) since the effects of these on equipment are, as a rule, far more damaging than those of residual nuclear radiation, which is the only significant delayed output. Nevertheless, as we saw in Chapter 6, there are some classes of electronic equipment which, even though they are able to survive the immediate effects, may have their performance seriously degraded by the gamma-ray dose rates from fallout. Where appropriate, the survivability criteria should specify the residual radiation dose rate which the equipment is to survive, and for how long it must operate at this dose rate.

The relative effectiveness against man of each of the weapon outputs varies with yield. As we have seen in previous chapters, for low-yield weapons immediate nuclear radiation is the ruling effect, that is, it produces casualties out to the greatest range from GZ. For intermediate yields blast usually takes over as the ruling effect, and for higher yields thermal radiation becomes dominant. Since an equipment and the men who are to operate it could be exposed to any of a variety of yields, we must begin by deciding on the range of yields to which the men and equipment might be exposed. This is called the threat yield spectrum. It is chosen after intelligence assessments have been made of the yields available to potential adversaries and how they might use them.

The nuclear environment in which the equipment is to survive is defined by the values of each effect which incapacitate 50 per cent of the operators. Here for simplicity we err on the safe side by assuming that there are no synergistic effects on man, that is, each output acts independently. Approximate values are:

Thermal radiation: for men in the open wearing full NBC clothing, as

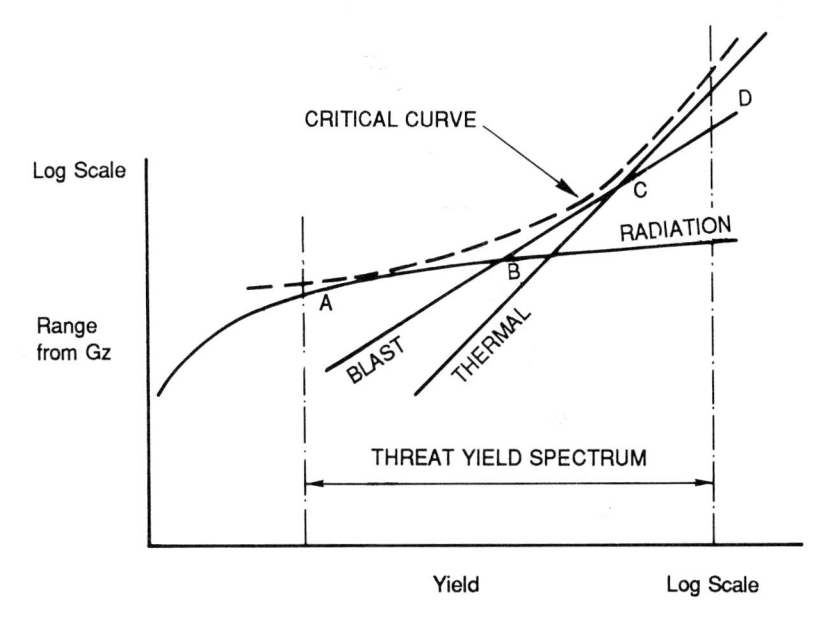

FIG. 8.1 Critical effects curve

shown in the frontispiece, a total thermal energy of 1.3 MJ/m^2 (30 cal/cm^2), is assumed to cause 50 per cent casualties. Men are deemed to be partially or completely protected against thermal effects if they are inside vehicles.

Blast: the blast which causes 50 per cent incidence of incapacitation to prone personnel who are picked up by the blast wind and subsequently strike a solid object. If men are protected against blast effects, the level is that which causes a 50 per cent probability of moderate damage to the means of protection, be it a vehicle or a shelter. For vehicles this is taken to mean the overturning of the vehicle.

Nuclear radiation: a total dose to the whole body of 2,600 cGy.

To find the ruling effect for any yield within the threat spectrum, and to derive survivability criteria, graphs as in Figure 8.1 are drawn. They show how the range at which men in a particular situation (e.g., in the open or inside a tank) suffer 50 per cent casualties from each of the weapon outputs varies with yield. The individual curves are called isocasualty curves. If we were concerned with men in infantry armoured vehicles or tanks there would be no thermal curve because we assume that they are completely protected against thermal effects. We also mark the limits of the threat yield spectrum on the graphs. The example in the Figure is typical in that it shows nuclear radiation as the ruling effect for the lowest yields, blast for intermediate yields, and thermal radiation for the highest.

Critical Effects Curve

The envelope of the isocasualty curves, shown by the dashed curve in the Figure, is called the critical effects curve. It defines the range for each yield within the threat spectrum at which a nuclear-hard equipment is required to have a high probability of surviving, since casualties to its operators at that range are expected to be 50 per cent. We could simply give this to the designer and leave it to him to work out the details of the thermal, blast and nuclear radiation levels his equipment must survive. However, to save him this effort it is standard practice to give him a set of nuclear survivability criteria, which is a detailed group of figures defining the severest levels of each effect which occur along the length of the critical effects curve.

If the equipment is to be used by men in the open the portion AB of the curve would correspond to a total radiation dose of 2,600 cGy but, as we saw in Chapter 6, at B a larger fraction of this would be from gamma rays than at A. For neutrons the reverse would apply. Because of the different effects which neutrons and gamma rays induce in electronics (but not in men), we must specify the maximum dose of each, which would be at A for neutrons and B for gamma rays. The designer must also be told the peak gamma-ray dose rate which, although it is irrelevant to the effects on man, is often crucial in determining how electronics will be affected.

The blast criteria describe the blast environment along the portion BC of the curve, but although this corresponds to a constant effect, such as the translation of prone personnel or the overturning of their vehicle, the blast parameters will vary along BC. Overpressures, both static and dynamic, will be highest at B, but the positive phase duration will be longest at C. Therefore two sets of criteria are needed, describing the blast wave at B and at C. Each must include the peak static and the dynamic overpressure, their durations and their impulses.

Similar considerations apply to the thermal criteria, which are taken from the portion CD of the critical effects curve. The greatest total energy will be at D, but the highest peak irradiance will be found at C, so again two sets of criteria are needed. Each set also includes the time to the second maximum of the thermal pulse, and a measure of its total duration. For both blast and thermal radiation it is up to the designer to decide which of the two sets of criteria presents the greater threat to each component of his equipment, and to design and test accordingly.

If an equipment is to be used on the exterior of a vehicle which houses its operators and protects them from thermal radiation, there will be no thermal isocasualty curve. In this case, derived thermal criteria would be produced; these would be the greatest values of the thermal parameters occurring along a critical effects curve formed by the nuclear radiation and the blast isocasualty curves.

Finally, for EMP we noted in Chapter 7 that there are standard pulse forms,

Nuclear Survivability

which are independent of both range and weapon yield, for the endo- and the exo-atmospheric pulse. Where they are appropriate, descriptions of one or both of these are included in the criteria. They cover rise times, peak fields, pulse duration and polarization for the electric and the magnetic field.

Standard Cases

From what has been said so far one might suppose that for every new Staff Requirement one would have to start from scratch by drawing a set of appropriate isocasualty curves and from them deriving a set of criteria unique to the equipment. If this approach were adopted there would exist a large number of sets of criteria, many of which would be very similar. We can simplify the situation. Almost all military equipment is operated on the battlefield by men. They are either in the open (i.e., unprotected) or else protected to some extent by a vehicle or shelter of some sort. The equipment may form part of, or be carried on or within, the vehicle or shelter. The vehicle or shelter will give the operators some degree of protection, depending on its type, against some of the weapon effects. Military vehicles fall into three broad categories:

wheeled vehicles,

light armoured fighting vehicles (AFVs), e.g., armoured personnel carriers (APCs) and infantry armoured vehicles (IAVs)

tanks and other heavy AFVs.

Within each category the response to each weapon output, and the protection provided against each output, is broadly similar.

It is therefore possible to relate the nuclear environment of the equipment to the environment of the operators as created by their vehicle, or lack of one, and to classify these environments into a small number of standard cases. We then only need to identify the most stringent of these environments in which the equipment will be used, and specify the appropriate standard case criteria. The standard cases are:

Case 1: Equipment associated with men in the open
The men are assumed to be wearing NBC protective clothing over full combat clothing, which gives them some protection against thermal, but none against blast or immediate nuclear radiation.

Case 2: Equipment associated with men in wheeled vehicles
The men are assumed to be protected from thermal effects by the vehicle, as almost any cover will give significant protection. The immediate radiation dose will be the same as for men in the open (2,600 cGy), since a typical wheeled vehicle affords little or no protection against neutrons and immediate gamma rays.

Case 3: Equipment associated with men in light AFVs and IAVs
Men are assumed to be protected against thermal and direct blast effects. The armour of some vehicles is too thin to provide significant protection against immediate nuclear radiation, so the radiation criterion, for both internal and external equipment, will be 2,600 cGy. In vehicles of more recent origin with heavier armour, the radiation criterion for internal equipment would, of course, still be 2,600 cGy, but for externally-mounted equipment it would be somewhat higher.

Case 4: Equipment associated with men in heavy AFVs
This case applies only to main battle tanks and their derivatives, which are assumed to give complete protection to their crews against thermal and direct blast effects, and a degree of protection against nuclear radiation. Radiation criteria are 2,600 cGy for internal equipment, but in the range 8,000–10,000 cGy for externally-mounted equipment (the value depends on the ratio of gamma to neutron dose, which is influenced not only by the transmission factors of the armour, but also by the threat yield spectrum selected).

Case 5: Equipment in supply dumps
A set of criteria which are not linked to the survival of men are derived for this case.

Case 6: Exo-atmospheric EMP only
Man is not affected by EMP. Because the exo-atmospheric EMP vastly out-ranges all other weapon effects, equipment can be damaged by it without men being at risk. There are equipments and systems for which full, balanced, nuclear-survivability criteria would be inappropriate or uneconomic, or sometimes impossible to meet for various fundamental reasons. In these cases hardening only against exo-atmospheric EMP is specified. Examples are home-based communications systems which are vital to post-strike operations but which are far removed from the theatre of operations.

Most Army equipments are covered by one of these standard cases, but there are a number of exceptions. Some items such as antitank mines and unattended ground sensors, once emplaced, do not have an operator nearby to whose survival their survivability can be linked. Other equipments may be intended for mounting on new classes of vehicle which are less vulnerable to blast, or are better protected against nuclear radiation, than their predecessors. Cases such as these must be considered individually; usually it is possible to find a logical basis for the derivation of a set of criteria.

Criteria for Shipborne and Airborne Equipment

Nuclear survivability is not exclusive to land-based forces; it must also be considered for equipment and systems intended for use in ships and aircraft. However, both ships and aircraft tend to be less resistant to weapon effects, particularly blast, than are army vehicles, or indeed men in the open. Furthermore, little can be done to improve their blast resistance without imposing unacceptable cost and weight penalties as well as performance degradation. In addition, it is unlikely that an aircraft pilot receiving a radiation dose of 2,600 cGy would be able to complete his mission. For these reasons criteria for ships and aircraft tend to be lower than the values for the Army.

The Designer's Approach to Nuclear Survivability

Once the designer or design team has been given the requirement from the Operational Requirements Branch a project management team will take on the assignment according to the requirement schedule and in-service date. There may or may not be a nuclear focus for the project, but in the UK most advice will come from the AWE. This is best sought at an early stage to avoid a first vulnerable design or the use of components which would be difficult to protect or harden later (retrospective hardening is usually expensive and sometimes impossible).

Designers should approach survivability with a flexible attitude; it may be that for economy or other reasons there have to be trade-offs on some criteria which are expensive or impossible to meet. Therefore a Project Manager must not be too rigid in his approach to nuclear survivability. He must also bear in mind that technology and materials are constantly changing and advancing, and that components used in one production run may have different intrinsic qualities and characteristics from those used in later runs, while still serving the same function.

There is no doubt that the designer has a difficult task in achieving total system hardening, especially for complex electronic communications and surveillance systems. He has four main effects to consider, each presenting its own unique set of problems. These are briefly summarized below.

Thermal Radiation

In Chapter 4 we stated that thermal effects are regarded as a bonus, and are not taken into account as a casualty producer. When it comes to our own equipment the reverse is true, though usually it is only the exposed surfaces which are affected. There is not a great deal of readily available information to draw on, so materials and finishes should be selected only after their response to thermal radiation has been tested in simulators (which we discuss in the next chapter). Fortunately, simulation is simple, quick and cheap. It is important to note that many plastics melt or ignite at quite low thermal levels, especially if they are thin.

Blast

There is a great deal of information on blast damage mechanisms from nuclear-weapon trials and from more recent large-scale, high-explosive detonations. Where it is available this should be drawn upon to avoid pitfalls. In essence, it is a problem of strength versus mobility in many cases, with the possibility of damage to appliqué and ancillary equipment as an added variable. There is also the synergistic effect of the thermal pulse which precedes the blast; this may lead to blast effects which exceed expectations. As with all the effects, those with little experience should consult experts at an early stage, and make use of simulation techniques if they are available.

Nuclear Radiation (TREE)

Unlike blast, which is principally a 'whole system' effect, TREE is basically a component effect. Usually it is impossible to mitigate the effects by radiation shielding, so the designer must select components which are known to be hard. Where this is not possible he must overdesign so that some degree of degradation will not impair the overall performance, or else use circumvention techniques. Simulator tests both on components and on subsystems may be needed if the latest device technologies are being used.

EMP

This, like blast, is a 'whole system' effect which is fairly well documented. It is essential to consider it early, in parallel with other electromagnetic problems, such as lightning, radiofrequency interference and TEMPEST, if one is to arrive at the simplest and cheapest solution. A good deal of guidance for the designer was given in Chapter 7. Poor maintenance or ill-considered modifications may easily convert an EMP-hard system into one which is highly susceptible. Maintenance of EMP hardness therefore poses more problems than do blast and thermal and nuclear radition, since it demands not only care by the designer, but also knowledge and vigilance from those responsible for in-service management and from those who use the equipment in the field.

Modifications and Maintenance

The survivability of equipment may be affected by inadequate maintenance, or by modifications made without proper consideration of the implications for nuclear survivability. This is particularly the case for EMP hardness, and to a lesser extent for TREE, but hardness to blast and thermal radiation may also be affected. Therefore it is essential that those responsible for modifications should have not only a good understanding of nuclear-weapon effects, but also an awareness of the design philosophy employed by those who originally developed the equipment.

Simulation

We have mentioned in this chapter the need for the simulation of the nuclear environment at various stages during development. The simulation techniques most widely available are summarised in Chapter 9.

Conclusion

In this chapter we have looked at weapon effects from the defensive viewpoint. This is just as important as the offensive aspects we have considered in earlier chapters. Nuclear survivability is not easy to achieve, and it is the golden rule that it must be addressed early in a project since post-design hardening is almost always difficult and expensive. Most new requirements contain specific criteria which the designer must try to satisfy. It will be at our peril that we neglect this aspect of any new equipment for battlefield or strategic use.

9. Vulnerability Assessment

Introduction

The goal of nuclear survivability cannot be achieved without practical tests to demonstrate the levels of the nuclear weapon effects an equipment can withstand. Theoretical studies are a help, but are rarely sufficient on their own.

In the 1950s many of the atmospheric weapon tests which were carried out were 'target response' tests, designed to establish the vulnerability of many types of military and civil equipment and structure. This was before the widespread use of solid-state devices in electronics; virtually all our electronic equipment was valve-based and so was little affected by nuclear radiation or EMP. The main interests at this time were the effects of blast and thermal radiation on equipment, and the effectiveness of armoured vehicles and other structures as nuclear radiation shields for their occupants.

Since the signing of the atmospheric test ban agreement in 1963, no further tests of this type have been carried out by the signatories. The agreement roughly coincided with the replacement of valves by semiconductor devices, and with the recognition of the threats to these devices which EMP and TREE present. Two alternatives to atmospheric tests are available: they are underground nuclear tests and simulations of the weapon environment.

Underground tests have some serious drawbacks. First, the necessary boreholes and horizontal tunnels to house the weapon and the equipments to be exposed are expensive to construct. Secondly, only a limited number of equipments or components can be exposed. Thirdly, the presence of rock around the tunnels significantly modifies some of the weapon outputs. For these reasons one must nowadays simulate the weapon environment to demonstrate equipment survivability.

There is no realistic way of simulating all the effects simultaneously, so that blast, thermal, gamma rays, neutrons and EMP arrive at an equipment in the right sequence and with the correct intensities, as illustrated in Figure 3.4. However, there is usually no need for this since, as a rule, each weapon output produces its own effects regardless of whether or not the other outputs are present. Therefore it is normal to expose equipment to only one output at a time. The exceptions are blast and thermal, the synergistic effects of which were mentioned in Chapter 4. The thermal radiation arrives first, and may weaken a structure, rendering it more vulnerable to the blast wave which arrives a second or two later. The sections which follow describe the simulation methods which are most widely used.

Blast Simulation

The three principal methods of simulating the blast wave in air are discussed below. They are usually only able to produce a blast wave of short positive phase duration, such as that from a yield of a few Kilotons at most. One must then use theory to predict the effects of the blast wave of much longer duration which a high-yield burst would produce.

Shock Tubes

A shock tube consists of a long tube divided into two sections by a diaphragm. One section is filled with compressed air; when the diaphragm is ruptured a shock wave travels down the other section where the equipment under test is installed. They are usually (but not always) quite small, and may then only be suitable for tests on small-scale models. For most equipments, therefore, we have to turn to the principal method of blast simulation – the blast tunnel.

Blast Tunnels

A blast tunnel consists of a long, conical tube like that shown in Figure 9.1. Its length may be hundreds of metres. When a small charge is detonated inside it, at its apex, a blast wave travels down the tube. Since the blast energy is confined by the tube instead of spreading freely in three dimensions, the blast from quite a modest amount of HE is similar to that from a low-yield nuclear burst.

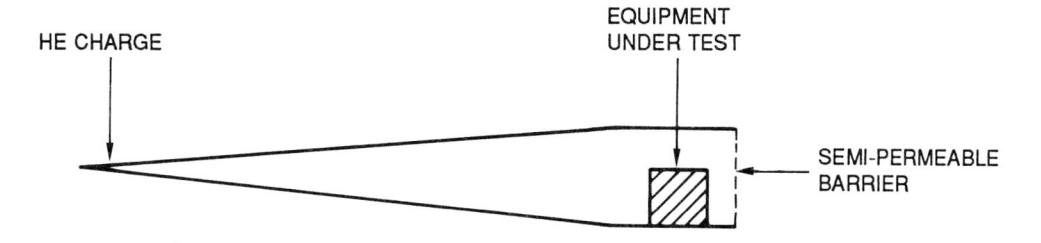

FIG. 9.1 Blast tunnel

The right-hand end must not remain open, nor must it be closed by a rigid barrier, since either of these would cause a reflected wave to move back down the tunnel, complicating the response of the target equipment. By placing across the end a series of slats or metal tubes, which act as a semi-permeable barrier, the reflected wave can be eliminated.

Equipment to be tested is placed in a 'working section' near the right-hand end. It may have to be tethered to prevent it from overturning and so damaging the tunnel and its installations. High-speed photography and strain gauging are among the techniques used to assess the response of the equipment under test. Since in blast-tunnel trials the equipment must usually be tethered, there is no indication of the damage it would suffer were it to topple or overturn. For this reason, equipments which have been tested in a blast tunnel are sometimes subsequently exposed to the blast from a large HE charge, which we describe next.

Large HE Charges

Every year or so the USA carries out a blast trial involving the detonation of a large explosive charge, usually several hundred tons of TNT or ANFO. For practical reasons the charges are usually surface bursts. Many equipments and structures, originating from a number of countries, are deployed at each of these trials.

Thermal Simulation

Realistic thermal simulation is not easy. Some of the simpler methods can irradiate only a very small area, perhaps only a few square centimetres. Most have a rather low colour temperature. This is the temperature of a hot solid which would give a similar distribution of thermal radiation over the ultraviolet, visible and infrared regions of the spectrum. The average colour temperature of the nuclear fireball is about 6000 K; for most simulators it is below this, so they are deficient in ultraviolet and visible when compared with a weapon.

Solar Furnaces

A solar furnace with a collection area of 30 m^2 can produce an irradiance of about 4 MJ/m^2 sec on a 10-cm^2 target. They have been used in the USA and France, but although their colour temperature is good they have not found much favour in the UK.

Arc and Strip Lamps

The simplest way of simulating a weapon's thermal pulse is by use of a xenon arc lamp, the energy from which is focused by mirrors on to a small area of a sample – a few square centimetres at most. About 20 MJ/m^2 can be delivered in a few seconds. With a colour temperature of 5600 K the thermal radiation is quite a good match with a nuclear weapon, but the small area irradiated means that the tests are somewhat limited in value. For example, it is difficult to decide how a given amount of thermal radiation affects the strength of a load-securing strap if one is able to irradiate an area of it only the size of a postage stamp.

Better in this respect is an array of tungsten-halogen strip lamps, which can deliver an irradiance of about 1 MJ/m^2 sec to a sample area of a few hundred square centimetres. The colour temperature is a rather modest 3700 K, but filters can be used to improve this somewhat.

Liquid Oxygen Simulators

From the point of view of the area illuminated, liquid oxygen thermal simulators provide by far the best method of thermal testing. They involve the combustion of aluminium powder in a number of jets of liquid oxygen. Unlike any of the other methods, they can deliver a useful thermal energy to a whole vehicle, but the colour temperature is rather low (less than 3000 K). They are often used in conjunction with blast simulation methods to reveal any synergistic effects; their particular merit is that they may well reveal thermal vulnerabilities which otherwise would be missed.

Nuclear Radiation

We saw in Chapter 6 that neutrons and gamma rays produce quite different effects on electronics. We also saw that the performance of a nuclear radiation shield against gamma rays gives us little indication of its performance against neutrons. Therefore sources of both neutrons and gamma rays are needed if we are concerned with TREE effects or with shielding.

Neutrons

As far as the neutron energy spectrum, the time to deliver a neutron dose, and the sample volume irradiated are concerned the neutron source which is nearest to the ideal is a fast pulsed reactor. This is like a sluggish nuclear weapon with a system of control mechanisms to terminate the pulse. Small ones can irradiate a single, small equipment with the required neutron dose; larger ones can cope with bigger units, or even a complete vehicle for neutron-shielding studies.

Thermal reactors, both steady state and pulsed, are more widely available, but their neutron energy spectrum is much less well matched to the weapon due to the influence of the moderator. The sample volume is limited, and, as a rule, only single devices or perhaps a printed circuit board can be accommodated. Steady state thermal reactors may take as long as several minutes to deliver the required neutron dose. During this time the annealing of neutron damage in semiconductors occurs, so a rather optimistic view of the degradation is obtained.

The radioisotope californium-252, which undergoes spontaneous fission, provides a simple yet effective source of neutrons for some shielding studies.

Gamma Rays

The most important gamma ray effect on materials is the generation of photocurrents in semiconductor devices. Since the current depends on the gamma dose rate, we need simulators which can provide the very high dose rates which are a feature of the gamma pulse from a weapon. No other source can give us a gamma dose rate anything like that from a weapon, so we have to use X-rays. These are identical to gamma rays in their properties (both are very energetic photons of electromagnetic radiation); they differ only in their origins.

Machines called flash X-ray sets are available consisting of a high-voltage pulser which delivers a very short pulse of several million volts to an X-ray tube. They can deliver the required dose rate to a small, complete equipment. However, it is not the photons themselves which produce effects on electronics – it is the Compton electrons ejected by the photons which are responsible for the copious production of electron-hole pairs. We can therefore study TREE phenomena in devices by exposing them to an intense pulse of electrons from a linear accelerator; but there are limitations to this approach. The penetrating power of electrons is much less than that of X- or gamma rays, and the electron beam from a typical linear accelerator is rather small in cross-section. Consequently we can irradiate only individual semiconductor devices, not complete circuits or equipments.

When we are interested in effects which depend on the total gamma dose rather than the dose rate a simpler approach is possible. Radioisotope sources which can deliver doses of a few thousand centigrays to devices or complete equipments in a few minutes are fairly widely available. The radioisotopes used are usually cobalt-60 or caesium-137; these sources are regularly employed for the sterilisation of medical equipment, food preservation, and the modification of polymers.

Similar but smaller sources may be used to simulate residual gamma radiation or for gamma ray-shielding studies; but the gamma rays they emit are somewhat lower in energy than weapon gammas. They are therefore more rapidly attenuated in a shield, so the results must be corrected to compensate for this.

Electromagnetic Pulse

The way in which it couples into long cables makes EMP the most difficult of the nuclear weapon effects to simulate realistically. Different approaches are needed depending on the size of the system, but all suffer from limitations, so that results from simulators need to be supported by detailed theoretical analysis.

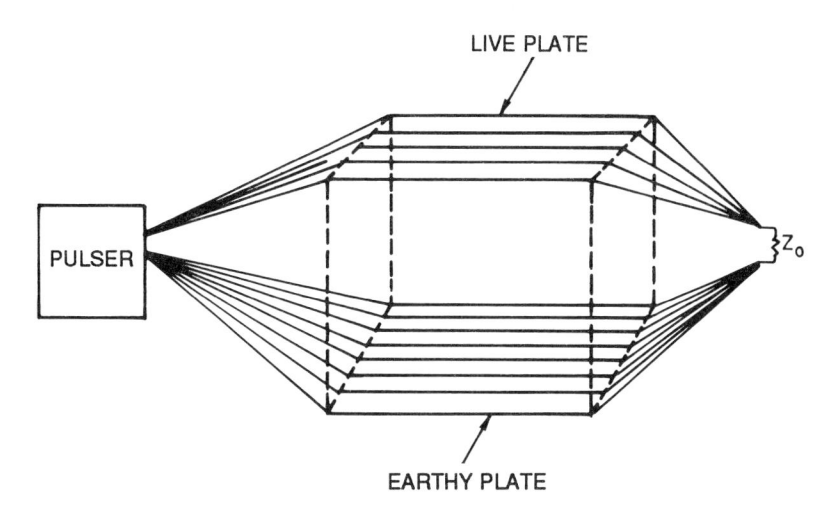

FIG. 9.2 Bounded wave EMP simulator

Bounded Wave Simulators

These, which are also called 'parallel plate' or 'guided wave' simulators, are convenient for tests on small systems. They comprise (Figure 9.2) a parallel-plate transmission line consisting of two sets of wires. Tapered sections connect them to a high-voltage pulser at one end and a resistive load (to prevent reflections) at the other. The plate separation is usually between 0.5 and 10 metres, though larger models have been built. Usually one of the plates is on the ground, so the electric field between the plates is vertically polarized.

In the pulser a power supply charges a capacitor which is discharged into the line. The peak electric field is simply the capacitor voltage divided by the plate separation, so a large simulator may need a charging voltage of half a million volts to give a 50 kV/m field. It is not too difficult to simulate the early part of the exo-atmospheric pulse, but simulation of the later parts, or of the endo-atmospheric pulse, is difficult or impossible. In any case, it would be pointless, since the comparatively small systems which can be accommodated in this type of simulator would not pick up much energy from these more slowly varying fields.

Bounded wave simulators are widely employed to assess the EMP survivability of small systems such as radios, vehicles and missiles, but some larger ones have been built to accommodate ships and aircraft. Being essentially a transmission line they do not radiate and thus do not create too many problems in the surrounding area. Their principal disadvantages are their vertical polarization (the real exo-atmospheric pulse is horizontally polarized) and the limited size of the system they can accommodate. Both of these may be overcome by the radiating simulators described next.

FIG. 9.3 Large, barge-mounted, radiating simulator for EMP assessment of ships
(*US Navy and Maxwell Laboratories*)

Radiating Simulators

These have a pulser like that used in the bounded wave simulator, but now it feeds either a vertical conical antenna (Figure 9.3) to give a vertically-polarized radiated field, or a horizontal bicone for horizontal polarization. Although they can cope with systems too large to be tested in a bounded wave simulator, they have their limitations. First, the field strength falls with distance from the antenna, so parts of a system may be exposed to fields below those of the EMP threat. Secondly, the radiated field may cause upsets or damage to nearby military or civil systems, so they are not popular as neighbours. Also, although there is not much evidence that EMP-like fields are harmful to man, animals or vegetation, their use concerns some environmentalists.

Current Injection

As we saw in Chapter 7, EMP interacts with the whole of a system, being picked up particularly by long cables. Some systems are far too large to be illuminated by any conceivable simulator: examples are extensive military and civil communication systems, and national grids for electric power distribution. In cases such as these the only practicable approach is to estimate the current waveform likely to be induced in each cable by EMP. A current pulse of this wave form is then injected into the equipment at the end of the cable to check that the protective devices installed there function as intended and protect the equipment from damage.

Shielding Measurements

For small systems with a global shield, testing with one of the simulator types described above reveals any imperfections of the shield. This approach may not be possible for a large shield, such as one protecting a static site. Then one must depend on careful supervision and testing during shield construction, current injection tests on incoming cables, and thereafter regular shielding tests on potential weak points, such as doors, hatches, and air inlets. Portable instruments for localised measurements like these are available, but access to both sides of the shield is needed.

Conclusion

Military equipments must be designed so that they will perform satisfactorily in many different environments. They have to survive high and low temperatures, lightning strikes, vibration and shock, electromagnetic emissions from nearby radar sets and so on. Nuclear survivability criteria could be viewed as just another set of requirements for the designer to satisfy. However, it is essential that we look at them rather differently.

If, due to deficiencies in design and inadequate testing, an equipment goes into service with insufficient protection against, say, vibration or lightning, these deficiencies are likely to show themselves fairly early in its service life. Steps can then be taken to improve its survivability in these respects.

The situation is quite different for the nuclear environment of blast, thermal radiation, nuclear radiation and perhaps most importantly, EMP. Equipment vulnerability to any of these is unlikely to reveal itself until nuclear weapons are used in earnest. Then it is too late: comprehensive testing during development, with simulated nuclear environments, is vital.

Bibliography

The books listed below are a sample of those which contain further information. The bold figures after each item refer to the chapters in this book, the contents of which are amplified in the cited publication.

S. Glasstone and P. J. Dolan, *The Effects of Nuclear Weapons* (US Dept. of Defense/Dept. of Energy, 3rd edn 1977, 1983); **1–7**

G. A. Jones, *The Properties of Nuclei* (Clarendon Press, Oxford, 2nd edn 1987); **1**

Richard Rhodes, *The Making of the Atomic Bomb* (Simon & Schuster, 1986; Penguin Booler, 1988); **1, 2**

R. L. Murray, *Nuclear Energy* (Pergamon, 3rd edn 1988); **1, 2, 6**

D. J. Bennett and J. R. Thomson, *The Elements of Nuclear Power* (Longman/Wiley, 3rd edn 1989); **1, 2, 6**

R. Serber, *The Los Alamos Primer* (University of California Press, 1992); **2**

S. T. Cohen, *The Neutron Bomb; Political Technological and Military Issues* (Institute for Foreign Policy Analysis, Cambridge, MA, 1978); **2**

F. Winterberg, *The Physical Principles of Thermonuclear Explosive Devices* (Fusion Energy Foundation, NY, 1981); **2**

N. J. Rudie, *Principles and Techniques of Radiation Hardening*, Vol. 1. *Thermal and Blast*; Vol. 2: *Nuclear Radiation*; Vol. 3; *EMP* (Western Periodical Co, CA, 2nd end 1980); **2–7**

British Medical Association, *The Medical Effects of Nuclear Weapons* (Wiley, 1983); **3–6**

G. C. Messenger and M. S. Ash, *The Effects of Radiation on Electronic Systems* (Van Nostrand Reinhold, 1986); **6**

R. Sherman, *EMP: Engineering and Design Principles* (Bell Telephone Laboratories, Whippany, NY, 1975); **7**

R. N. Ghose, *EMP Environment and System Hardness Design* (Don White Consultants, Gannesville, VA, 1984); **7**

Glossary

Absorbed dose
: energy absorbed per kg by an irradiated material; unit: centigray

Absorbed dose rate
: energy per kg absorbed in unit time, usually in cGy/hour

Activity
: rate of decay of a radioactive material, in disintegrations per second (becquerels)

Alpha decay
: radioactive decay mode of some heavy atoms, involving alpha particle emission

Alpha particle
: particle (2 protons, 2 neutrons) emitted in alpha decay

Air burst
: nuclear explosion at a height such that the fireball does not touch the ground

Ambient pressure
: normal air pressure (101 kPa or 14 psi)

Ampere per metre
: unit of magnetic field strength

ANFO
: Ammonium nitrate/fuel oil (used as an explosive)

Atomic number
: total number Z of protons, or electrons, in an atom

AWE
: Atomic Weapons Establishment

Balanced survivability
: ability of equipment to survive all nuclear weapon effects

Base surge
: radioactive droplet cloud produced by an underwater burst

Beta burns
: damage to the skin caused by beta particles

Beta decay
: radioactive decay mode, shown by neutron-rich nuclei, involving emission of an energetic electron

Beta particle
: electron emitted in beta decay

Binding energy
: energy required to split a nucleus into its constituents

Binding energy per nucleon
: binding energy divided by number of nucleons, a measure of stability

Blast tunnel
: conical tunnel used to simulate blast from a nuclear weapon

Blast wind
: air movement associated with blast wave (source of the dynamic pressure)

Boosted fission weapon
: weapon with a small fusion stage which increases the efficiency of fission stage

Calorie
: a non-SI energy unit (= 4.186 joules)

Centigray (cGy)
: military unit of absorbed radiation dose

Centigray/hour
: military unit of absorbed radiation dose rate

Chain reaction
: a self-sustaining chemical or nuclear reaction

Chemical prophylaxis
: administration of drugs to counteract radiation effects

Circumvention
: prevention of corruption in electronics by choice of system configuration or operational procedures

Clean weapon
: weapon obtaining most of its energy from fusion, resulting in little residual radiation.

Collateral damage
: unwanted side effects arising from use of nuclear weapons

Colour temperature
: temperature of a heat source, estimated from the spectrum of the radiation it emits

Glossary

Compton current
: current consisting of electrons ejected by Compton scattering

Compton electron
: electron ejected by Compton scattering

Compton scattering
: interaction of a gamma ray photon with an atom, resulting in ejection of an energetic electron

Contamination
: loose material (radioactive fallout on man, equipment or ground

Contamination meter
: device for detecting beta particles emitted by radioactive contamination

Coupling
: energy transfer from EM waves (q.v.) to a conductor, or between conductors

Criteria
: levels of nuclear-weapon effects which an equipment is required to survive

Critical
: state of a quantity of fissile material, in which a fission chain reaction is just able to maintain itself

Critical mass, size
: mass or size of a quantity of fissile material in which a fission chain reaction is just able to maintain itself

Current injection
: method of testing EMP (q.v) survivability by injecting into equipment currents such as those EMP would induce

Decay
: breakup of unstable (radioactive) nucleus

Depleted uranium
: uranium with less than the normal 0.7 per cent of U-235

Deposition region
: region round a weapon burst in which gamma rays deposit their energy

Deuterium
: hydrogen isotope $^{2}_{1}H$; 0.013 per cent of natural hydrogen

Diffraction
: the spreading of a (blast) wave round a target

Diffraction loading
: translational force on a target arising from static overpressure

Disintegration
: emission of particles and gamma rays by an unstable nucleus

Displacement
: movement of an atom from its original position in a crystal lattice

Dose, dose rate
: *see* Absorbed dose, Absorbed dose rate

Dose equivalent
: measure of the effectiveness of radiation at producing late somatic effects; unit: sievert

Dose rate meter
: instrument which measures gamma ray dose rate from residual radiation

Dosimeter
: instrument which measures the total radiation dose received by the wearer

Drag loading
: translational force on a target due to dynamic pressure

Dynamic pressure
: pressure caused by blast wind

Early somatic effects
: radiation effects on man which appear from minutes to months after irradiation

Efficiency
: percentage of fissile material in a weapon which is actually fissioned

Elastic scattering
: interaction between a particle and a nucleus in which the kinetic energy of the particle is shared between them

Electromagnetic pulse
: intense pulse of radio waves emitted when a weapon is detonated

Electromagnetic radiation (waves)
: radiations comprising the electromagnetic spectrum: gamma rays, X-rays, ultraviolet, visible, infrared, and radio waves

Electrons
: light, negatively-charged particles which orbit the nuclei of atoms

Electron-volt (eV)
: energy unit used in atomic physics; $1\ eV = 1.6 \times 10^{-19}$ joule

EMP
: electromagnetic pulse

Glossary

EM waves
: electromagnetic waves

Endo-atmospheric
: within the atmosphere (below 30 km)

Enhanced radiation weapon
: low-yield weapon with most energy derived from fusion

Enriched uranium
: uranium with U-235 content above the normal 0.7 per cent

eV
: electron-volt (q. v.)

Excitation
: the raising of an atomic electron to a higher orbit

Fallout
: radioactive weapon debris, mainly fission products, deposited on the earth's surface

Fast reactor
: reactor with highly-enriched fuel and no moderator; fast neutrons cause the fissions

Film badge
: device using photographic emulsion to measure dose; used as a personal dosimeter

Fireball
: luminous sphere of hot gas formed by a nuclear explosion

Fissile material
: material which can be fissioned by neutrons (U-235, Pu-239)

Fission
: the splitting of a heavy nucleus into two roughly equal parts with release of much energy and several neutrons

Fission products
: the highly radioactive products of fission

Flash blindness
: the effects of thermal radiation on the eye (dazzle, retinal burns)

Flash X-ray set
: device for producing a short intense pulse of X-rays for the study of radiation effects

Fuel rod
: a cylinder of uranium for use in a reactor

Fusion
: the joining of two light nuclei (usually hydrogen isotopes) with a large energy release

Gamma rays
: energetic photons of electromagnetic radiation emitted in most radioactive decays and in other nuclear reactions

Genetic effects
: radiation effects on man involving harmful changes to the genes, and affecting later generations

Global shield
: a metal shield round a system to prevent entry of EMP (q.v.)

Governing effect
: see Ruling effect

Gray (Gy)
: unit of absorbed radiation dose

Ground burst
: burst on or close to the surface of the earth

Ground range
: distance from ground (or surface) zero to a target

Ground wave
: EM (radio) wave which follows the earth's surface

Ground zero (GZ)
: point on land surface immediately below a nuclear burst

Half-life
: the time in which half the atoms in a sample of radioactive material decay

Half-thickness
: the thickness of a specified material which reduces the intensity of X- or gamma rays by half

Hardening
: see Nuclear hardening

HE
: high explosive

Heavy hydrogen
: the hydrogen isotope $^{2}_{1}H$ (deuterium)

Heavy water
: water with deuterium in place of ordinary hydrogen; used as a moderator

HEMP
: high altitude (exo-atmospheric) electromagnetic pulse

Hertz (Hz)
: unit of frequency (1 Hz = 1 cycle/sec)

Glossary

High altitude burst
: burst at a height of more than 30 km

High yield weapon
: weapon with yield of about 100 kT or more

Hydrodynamic enhancement
: effect of blast, easing the passage of gamma rays through the air

Hydrogen bomb
: see Thermonuclear weapon

Immediate nuclear radiation
: radiation (neutrons and gamma rays) emitted within one minute of a burst

Implosion
: detonation of HE in such a way that it compresses fissile material and renders it supercritical (q.v.)

Initial nuclear radiation
: see Immediate nuclear radiation

Impulse
: the product of force (or pressure) and the time for which it acts

Inelastic scattering
: a process by which energetic neutrons lose energy to nuclei, which subsequently emit gamma rays

Ion
: an atom which, having lost one or more electrons, has a net positive charge

Ionization
: removal of an electron from an atom

Ionization chamber
: a device in which a current of ions provides a measure of radiation dose rate

Ionizing radiation
: see Nuclear radiation

Ionosphere
: region of the outer atmosphere, of height between 60 and 400 km, in which there is normally appreciable ionization

Irradiance
: rate of receipt of thermal energy, in joule/m^2/sec (W/m^2)

Isotopes
: forms of an element with differing numbers of neutrons in their nuclei

Joule, J
: unit of energy (thermal, kinetic, etc.)

K
: temperature in kelvins (temperature in degrees Celsius = temperature in K -237.15)

Kilo, k
: 1000, e.g., kJ, km, kPa, kHz, etc.

Kiloton (kT)
: unit of nuclear weapon yield (1 kT = 10^{12} calories = 4.186×10^{12} joules)

Late somatic effects
: radiation effects on man which do not become apparent for many years

Linear accelerator
: a device for accelerating electrons to high energy, used to simulate gamma effects on electronics, or to produce X-rays

Lithium deuteride
: lithium hydride with deuterium replacing ordinary hydrogen; used in high-yield weapons as fusion fuel

Local fallout
: intense fallout downwind from GZ (q.v.) resulting from a surface burst

Low air burst
: a nuclear weapon burst at a height of a few tens to a few hundred metres

Low-yield weapon
: weapon with yield in the range 0.1–10 kT

Mach stem
: intensified blast wave resulting from the merging of direct and ground-reflected waves

Mass number
: total number A of nucleons (protons and neutrons) in a nucleus

Mega, M
: 10^6, e.g., MJ, MHz, MeV, etc.

MeV
: million electron-volts (*see* eV)

Micro
: 10^{-6}

Micron
: 10^{-6} metre

Milli, m
: 10^{-3}, e.g., as in mm or mJ

Moderator
: a material containing deuterium, hydrogen or carbon which, by elastic scattering, slows down neutrons in a thermal reactor

Monitor
: instrument which measures gamma ray dose rate from residual radiation

Multiplication factor
: the ratio of the number of fissions in successive generations in a chain-reacting system

Mutation
: a change in the heredity information carried by the genes

Nano, n
: 10^{-9}, e.g., nanosecond (ns)

Natural radioactivity
: radioactive decay of naturally-occuring elements, e.g., uranium and thorium

Negative phase
: the portion of the blast wave where the pressure is below the ambient

NEMP
: nuclear electromagnetic pulse, i.e., EMP

Neutron
: uncharged particle present in all nuclei except hydrogen-1

Neutron bomb
: *see* Enhanced radiation weapon

Neutron induced activity (NIA)
: radioactivity induced in materials by capture of weapon neutrons

Newton (N)
: unit of force

Nuclear force
: force between nucleons, responsible for binding them together in the nucleus

Nuclear fission, fusion
: *see* Fission and Fusion

Nuclear hardening
: use of techniques which reduce the effects of nuclear weapons on equipment

Nuclear radiation
: particles and electromagnetic radiations which are energetic enough to ionize materials they encounter

Nuclear reactor
: a device in which a nuclear fission chain reaction is designed to occur

Nuclear survivability
: the ability of a system to continue to function after exposure to nuclear-weapon effects

Nucleon
: a particle in a nucleus (proton or neutron)

Nucleus
: the small dense core of an atom containing the nucleons

Optimum height of burst
: the burst height which maximizes the range of blast effects from a weapon

Overpressure
: air pressure excess over ambient pressure in a blast wave

Pascal (Pa)
: unit of pressure, equal to $1N/m^2$

Peak static overpressure (PSO)
: peak overpressure in a blast wave, at the shock front

Personal dosimeter
: instrument which indicates the total radiation dose received by the person wearing it

Photocurrent
: current which flows in a semiconductor device when exposed to high gamma ray dose rates

Photographic film badge
: *see* Film badge

Photons
: 'particles' of electromagnetic radiation with energy dependent on radiation type (gamma, X-ray, ultraviolet, visible light, etc.)

Plasma
: matter in a completely ionized state, with all atoms stripped of their electrons

Polarization
: direction of electric field in an EMP radio wave

Positive ion
: atom stripped of one or more electrons, so having a net positive charge

Positive phase
: time during which blast wave pressure exceeds the ambient value

Prompt gamma rays
: gamma rays emitted from a weapon during the fission chain reaction

Protection factor
: reciprocal of transmission factor (q.v.)

Proton
: positively-charged particle present in all nuclei

psi
: pounds per square inch; pressure unit, equal to 6.895 kPa

Pu-239
: fissile isotope of the artificial element plutonium

Pulsed reactor
: reactor held in a slightly supercritical state for a short time, to provide a pulse of fission neutrons

Pulser
: device for generating a high-voltage-pulse, to drive a flash X-ray set or EMP simulator

Quality factor
: relative effectiveness of a dose of radiation, eg neutrons, at causing late somatic damage, compared with gamma rays

Quartz fibre dosimeter
: personal dosimeter (q.v.) which measures absorbed dose from gamma rays

Rad
: former unit of absorbed dose, identical to the centigray

Radiation
: term used to describe all EM radiations, and also particulate radiations (alpha and beta particles and neutrons); in this book it is used to cover nuclear radiation

Radiation monitor
: *see* Dose rate meter

Radioactivity
: spontaneous decay of instable nuclei, usually by alpha or beta emission

Radioisotope
: radioactive isotope of an element, either natural or artificial

Reactor
: *see* Nuclear reactor

Residual nuclear radiation
: radiation emitted more than one minute after a burst (from fallout and NIA)

Retinal burns
: burns on the eye retina produced by thermal radiation focused on it by the lens

Radiophotoluminescence (RPL)
: increased fluorescence shown by some materials after irradiation; utilised in RPL dosimeter

Ruling effect
: the output of a particular weapon yield which produces damage or casualties out to the greatest range

Scaling laws
: laws which enable outputs or effects of any yield weapon to be deduced from those produced by 1 kT

Second-degree burns
: skin burns causing pain and blistering

Seven and ten law
: empirical law for the decay of fallout

Shield
: (a) material for absorbing nuclear radiation (gamma rays, neutrons), or (b) conducting structure around electronics to protect them from EMP

Shock front
: leading edge of the blast wave

Shock tube
: device using compressed air to simulate a blast wave

Shock wave
: *see* Blast wave

Sievert (Sv)
: unit of dose equivalent

Simulator
: device for simulating one of the nuclear weapon outputs

Sky wave
: radio wave reflected from the ionosphere

Slant range
: direct distance from an explosion to a target

Somatic effects
: biological effects of nuclear radiation on man – divided into early and late effects

Source region
: *see* Deposition region

Spontaneous fission
: fission of a nucleus without the intervention of a neutron

Standard case
: set of criteria applied to a whole class of similarly protected equipments

Steady state reactor
: reactor operating at constant power, producing a constant neutron dose rate

Subcritical
: fissile material with multiplication factor less than 1.0, so that a chain reaction dies away

Supercritical
: fissile material with multiplication factor greater than 1.0, so that a chain reaction develops

Surface burst
: weapon burst on or close to the surface of land or water

Surface cutoff
: cancellation effect caused by reflection at the surface of an underwater shock wave

Surface zero (SZ)
: point on the surface directly below (or above) a burst above (or below) water

Survey meter
: *see* Dose rate meter

Synergistic effects
: enhanced equipment vulnerability, due to combined effects of thermal radiation and blast

Tamper
: heavy material around the fissile material in a weapon to contain it for the time needed to give the desired yield

TEMPEST
: the inadvertent emission by equipment of EM waves carrying classified information which could be picked up by an enemy.

Thermal maximum, t_{max}
: time when the thermal output reaches its second maximum

Thermal minimum, t_{min}
: time when the thermal output dips to a minimum before peaking again at t_{max}

Thermal neutron
: neutron slowed to very low energy, usually by a moderator

Thermal radiation
: ultraviolet, visible and infrared radiation from a weapon

Thermal reactor
: reactor in which fissions are induced by thermal neutrons

Thermoluminescence dosimeter (TLD)
: device using light emission when certain irradiated materials are heated; used as a personal dosimeter

Thermonuclear fusion
: fusion of light nuclei induced by heating to an extremely high temperature

Thermonuclear weapon
: weapon employing a thermonuclear fusion stage

Threat yield spectrum
: the range of yields which an equipment may have to survive

Transient radiation
: immediate nuclear radiation

Transmission factor
: the fraction of an external nuclear radiation dose which is received within a shield

Transmissivity (T)
: the fraction of the thermal radiation which reaches a target, relative to that which it would receive if no atmosphere were present

Transuranic element
: elements not found in nature, with atomic numbers greater than 92, e.g., plutonium (atomic number 94)

TREE
: transient radiation effects on electronics.

Triple point
: the top of the Mach stem, where it intersects the direct and the ground-reflected blast wave

Tritium
: the hydrogen isotope, $^{3}_{1}H$, which is radioactive and does not occur naturally

U-235
: fissile isotope of uranium, 0.7 per cent of the natural element

U-238
: non-fissile isotope of uranium, 99.3 per cent of the natural element

Underpressure
: amount by which the pressure is below the ambient in the negative phase of the blast wave

Visibility
: range at which a large dark object can just be seen in daytime against the horizon sky

Volt per metre (V/m)
: unit of electric field strength

Weapon outputs
: the thermal radiation, blast, nuclear radiation and EMP produced by a weapon detonation

Waveguide below cutoff
: a narrow tube (or bundle of tubes) through which only the highest frequency EM waves can pass

Xenon arc furnace
: a device for producing thermal radiation similar to that from a weapon

X-rays
: energetic photons of EM radiation emitted by a weapon, or generated when fast electrons strike heavy elements

Yield
: the total energy output of a nuclear weapon, in kilotons (kT), denoted by W

Index

Activity 10
Alpha decay 8, 78
Alpha particle 8, 65
   range 65
   quality factor 71
Atomic number 3

Balanced survivability 107
Base surge 81
Beta burns 72
Beta decay 8, 13, 29, 78
Beta particle 9, 65
   range 29, 65, 71
Binding energy 7
Blast wave 25, 28, 47
   dynamic pressure 28, 50
   diffraction loading 55
   drag loading 55
   effect of ground 51
   effect of height of burst 28, 34, 52
   exo-atmospheric burst 32
   impulse 55
   Mach reflection 52
   negative phase 28, 50
   positive phase 28, 49
   positive phase duration 49
   reflection 51, 54
   scaling laws 52
   shock front 26, 28, 49
   static overpressure 48, 61
   subsurface burst 60
   surface burst 60
   underwater burst 61
   velocity 48, 61
Blast wave effects 56
   cratering 60
   on aircraft 59
   on buildings 59
   on forests 59
   on man 56
   on military equipment 56, 113
   on ships 59, 62
Blast protection 58
Blast simulation *see* Simulators
Boosted fission weapon 22

Cables, EMP effects 99, 102
Centigray 67
Chain reaction 13, 16
Charged particle interactions 65
   ranges 29, 65
Circumvention (TREE) 82
Colour temperature 40, 118
Compton current 92, 94
Compton scattering 65, 92, 94
Crater formation 60
Critical effects 110
Critical mass 15, 20
Current injection 122

Dazzle 41
Decontamination 87
Deuterium 6, 11, 21
Displacements 66, 82
Dose rate meters 84
Dose units 67
   absorbed dose 67
   dose equivalent 71
   quality factor 71
Dosimeters, neutron 84
   phosphate glass 84
   photographic film 84
   quartz fibre 83
   radiophotoluminescence 84
   thermoluminescence 84

Elastic scattering 66
Electromagnetic pulse (EMP) 25, 91
   endo-atmospheric 92
   exo-atmospheric 94
   magnetohydrodynamic 96

source region 92, 94
   system generated (SGEMP) 98
EMP coupling 99
   cables 99, 102, 104
   aircraft 97
EMP damage 99
EMP protection 100, 114
   earthing 104
   filters 103
   maintenance 103, 114
   modifications 103, 114
   precautions in use 103
   protective devices 103
   shielding 100, 123
EMP simulators *see* Simulators
Energy distribution in fission 32
Enhanced radiation weapon 23, 32
Enrichment 14, 21
Excitation 5

Fallout *see* Radiation, residual
Fibre optics 83, 103
Film badge 84
Filtration 87
Fireball development 25
Fission, energy release 12, 16
   neutron-induced 12
   neutron release 12
   spontaneous 20
Fission products 12, 29, 78
Flash blindness 41
Fusion 11, 17, 21

Gamma rays, emission 8, 9, 29, 73
   interactions 65
   shielding 65, 88
   spreading and absorption 29, 66
Gray 67
Ground bursts *see* Surface bursts
Ground zero 28
Gun-type weapon 16

Half-life 10
Half-thickness 66
Heat *see* Thermal radiation
Hydrodynamic enhancement 74

Image intensifiers 83
Immediate transient incapacitation 169
Impact bursts *see* Surface bursts
Implosion weapon 18

Inelastic scattering (neutrons) 67
Initiation, weapon 15, 17
Ionization 5
   chambers 83, 84
Ionosphere, radiation effects on 104
Irradiance (thermal) 40
Isocasualty curve 109
Isotope 6

Lithium deuteride 21

Mach stem 52
Man, blast effects 56, 109
   EMP effects 99, 122
   radiation effects 68
   thermal effects 40, 108
Mass number 3
Moderators 11, 19, 67
Multiplication factor 15

Natural radioactivity 8, 10
Neutron 3, 67
   capture 11, 67
   elastic scattering 66
   induced activity (NIA) 30, 80, 88
   inelastic scattering 67
   quality factor 71
   reactions 11
   thermal 12, 66
Neutron bomb 23, 42
Nuclear burst indicator 80
Nuclear forces 7
Nuclear radiation *see* Radiation
Nuclear reactions 11
Nuclear reactors 19, 119
Nucleon 7
Nucleus 3

Pascal 47
Photocurrent 82, 120
Plutonium 10, 19, 78
Portable dose rate meter 84
Propagation effects 104
Prophylaxis, chemical 90
Protection factor (nuclear radiation) 89
Proton 3, 66

Quality factor
Quartz fibre dosimeter

Index

Rad 107
Radiation *see* Alpha, Beta, Gamma, Neutron, X-rays
Radiation, immediate 30, 72
  from fission weapon 29, 73, 77
  from enhanced radiation weapon 32
  gamma rays 29, 73
  gamma dose 73
  gamma-dose rate 75, 81
  neutrons 29, 73
  neutron dose 73
Radiation, residual 30, 77, 80
  decay law 78
  decontamination 87
  dose to aircraft 80
  dose to ships 80
  effect of burst height 30, 77
  effect of weather 80
  fallout 30, 77
  filtration 87
  from air burst 30, 77
  from surface burst 31, 77
  from underwater burst 31, 80
  neutron-induced activity 30, 80, 88
  prediction 79
Radiation effects on man 68
  bone marrow 68
  cancers 70
  cell division 68
  central nervous system 69
  chemical prophylaxis 90
  gastro-intestinal tract 69
  genetic effects 70
  inhalation 70
  ingestion 70
  skin 72
  somatic 68
Radiation effects on material 81s
  electronics (TREE) 81, 114
  fibre optics 83
  glasses 82
  image intensifiers 83
  ionosphere 104
  plastics 83
  semiconductors 81
Radiation instrumentation 83
  contamination meters 86
  dose rate meters 84
  personal dosimeters 83
Radiation protection 86
  electronics 82, 86
  man 86
Radiation shielding 88
  against alpha particles 65
  against beta particles 65, 71
  against gamma rays 65, 88
  against neutrons 66, 89
Radiation shielding AFVs 89
Radiation simulation *see* Simulators
Radioactivity, natural 8, 10
  artificial 11
Radio propagation 104
Reactions, nuclear 11
Retinal burns 41

Scattering, elastic 66
  inelastic 67
Shielding *see* EMP, Radiation shielding
Shock front 26, 28, 49
Shock tubes 117
Sievert 71
Simulators 116
  blast 117
  EMP 120
  radiation 119
  thermal 118
Spark gaps 103
Slant range 37, 73
Surface bursts 31, 38, 60, 93
  cratering 31, 60
  fallout 31, 77, 80
  thermal radiation 31, 38
Survivability, airborne equipment 112
Survivability, army equipment 106
  balanced survivability 107
  criteria 110
  critical effects 110
  designer's approach 113
  policy 106
  project management 113
  ruling effects 33, 109
  staff targets 106
  standard cases 111
  threat yield spectrum 108
Survivability, shipborne equipment 112
Synergistic effects 42, 108, 114

Tamper 16, 19, 22
Thermal radiation 25
  characteristics 25, 36

colour temperature 40, 118
high-altitude bursts 32
irradiance 40
low air bursts 36
origin 25
reflection 38
scaling laws 37
second maximum 27, 39, 80
Stefan-Boltzmann Law 26
thermal maximum 27, 39
thermal minimum 27
transmissivity 37
Thermal radiation effects 40, 113
   eyes 41
   human skin 40
   materials 42, 113
   optical devices 44
Thermal protection 38
   equipment 45
   eyes 44
   optical devices 45
   skin 41, 44
Thermal simulators *see* Simulators
Translation of prone personnal 56
Transmission factor (radiation) 89
Transmissivity (thermal) 37

Triple point 52
Tritium 7, 11, 21

Underground bursts 31, 60, 77
Underwater bursts 31, 61, 80
   blast effects 31, 61
   radiation effects 31, 80
Uranium 3, 7, 14, 22

Visibility 37

Waveguide beyond cutoff 101
Weapon, boosted fission 22
   enhanced radiation 23
   gun-type 16
   high-yield 22
   implosion 18
   initiation 15, 17
Weapon accidents 81
Weapon effects *see* Blast, Thermal, etc.
Weapon outputs *see* Blast, Thermal, etc.
   time histories 34

Yield 16

X-rays 6, 23, 26, 32